The Return to Cosmology

STEPHEN TOULMIN

The Return
to
Cosmology

POSTMODERN SCIENCE AND
THE THEOLOGY OF NATURE

UNIVERSITY OF CALIFORNIA PRESS
Berkeley   Los Angeles   London

UNIVERSITY OF CALIFORNIA PRESS
Berkeley and Los Angeles, California

UNIVERSITY OF CALIFORNIA PRESS, LTD.
London, England

First Paperback Printing 1985
ISBN 0-520-05465-2

Library of Congress Cataloging in Publication Data

Toulmin, Stephen Edelston.
  The return to cosmology.

  1. Cosmology.   I. Title.
BD511.T68   113   82-40088
AACR2

PRINTED IN THE UNITED STATES OF AMERICA

1 2 3 4 5 6 7 8 9

Contents

Introduction: Thinking about the Universe   1

PART ONE: SCIENTIFIC MYTHOLOGY (1951–1957)

Scientific Theories and Scientific Myths   21

The Limits of Cosmology   33

Ethics and Cosmic Evolution   53

Science and Our View of the World   72

Conclusions   81

PART TWO: A CONSIDERATION OF COSMOLOGISTS (1964–1980)

Arthur Koestler I (1964)   89

Pierre Teilhard de Chardin (1965)   113

Arthur Koestler II (1968)   127

Jacques Monod (1971)   140

François Jacob (1974)   156

Carl Sagan (1977)   165

Arthur Koestler III (1979)   176

Gregory Bateson (1980)   201

PART THREE: THE FUTURE OF COSMOLOGY:
POSTMODERN SCIENCE AND NATURAL
RELIGION (1981)

All Cohaerence Gone   217

Death of the Spectator   237

The Fire and the Rose   255

Index   275

Introduction
Thinking about the Universe

I

This is a book about the fascinations and the frustrations of cosmology. Ever since human beings first began to reflect about, and to discuss, their situation within the world of natural things, their most comprehensive ambition has been to talk sense about *the Universe as a Whole*. In practical terms, this ambition has reflected the need to recognize where we stand in the world into which we have been born, to grasp our place in the scheme of things and to feel at home within it. In intellectual terms, meanwhile, it has stretched our powers of speculation and imagination like no other ambition, requiring us to extend the scope of our thoughts and our language beyond all natural boundaries, so that they become "all inclusive." Yet, again and again, the tasks so created have turned out to be ringed around with mines and booby traps. As the centuries have passed, accordingly, the problem of recognizing what kinds of statements, evidence, and arguments can lead us to sound conclusions about "the entire Universe"—that most mysterious object of thought—has become not easier, but harder.

Our cosmological ambitions have, in practice, too often deceived us into accepting fallacious or nonsensical inferences, incautious extrapolations, premature generalizations, or sheer confusions of category. The whole expanse of Space, for instance, is not just one more volume, which simply happens to be larger than all other volumes. Nor is the totality of Time just one more historical period, longer than all other periods of time, but otherwise *comme les autres.* So we cannot just extrapolate our familiar ideas about smaller regions of space and shorter periods of time, and apply them directly to Space and Time ''as wholes.'' Nor, for that matter, can we use our everyday discoveries about each and every limited, particular kind of thing as a secure foundation for conclusions about ''the All'' or ''the Whole.'' Whatever else is or is not true in our thinking about the universe, cosmology is a field in which we continually have to watch our steps; and we must look at cosmological arguments that have a strong initial appeal to us a second or third time—just *because* they were at first so appealing. For the speculative ambitions of cosmology have at all times been open to counterchallenge: perhaps, after all, the truth about the Universe as a Whole is unknowable to human beings, can be expressed in no human graphs, equations, or languages, and even eludes human thought and investigation completely.

Still, the fascination of these problems remains. From Thales and Anaximander to John Wheeler and Carl Sagan, the most imaginative and adventurous physicists have always conceived of the scope of physical science in comprehensive terms; from the first chapter of *Genesis* or the *Mahabharata* up to Engels's *Dialectics of Nature* and Teilhard de Chardin's *Phenomenon of Man*, the proper place of human affairs within the larger world of the natural creation has been the topic of cosmological theories and myths; and this entire lengthy debate has veered, unpredictably, between extremes of boldness and skepticism. The early philosophers of Ionia and Italy showed no hesitation in speculating that the entire Universe is basically constituted of some single underlying substance, whether Water or Breath or Mentality. But their bold claims at once provoked the skeptical criticism of Heraclitus, in whose eyes the contingent (or ''happenstance'') character of all human observations precluded any discovery of eternal, comprehensive truths about Nature, and later shipwrecked on the objections of Socrates, for whom the heavens were both practically and intellectually out of reach, and the only certainties were matters of direct human experience and humane concern. Socrates' own objections to the claims of natural philosophy, however, were countered in their turn by his pupil, Plato, who argued that the abstract theorems of geometry provide a model for scientific knowledge of reality which can transcend all the contingencies and other limitations of earlier pre-Socratic thinking. And once again, some two thousand years later, we find cosmological speculation repeating the same pendulum swings. The European rediscovery of the Greek scientific traditions at the time of the Renaissance stimulated, at first, the skeptical criticisms of Montaigne and his fellow humanists, only for these objections to be answered by the counterarguments of the new seventeenth-century "experimental and mathematical philosophers," notably Galileo Galilei and René Descartes.

So, from the outset, all attempts at finding a way to speak about the Nature of the Whole, without abandoning familiar human language, ran into philosophical difficulties of a peculiar kind. On the one hand, everyone involved in that cosmological search has felt a strong urge to press up against and, if necessary, even to overreach the limits of language and the intellect. On the other hand, a sense of the need for restraint, in philosophical as well as in practical matters, has also kept us aware of the risks and exaggerations to which that search exposes us, and so has held us back. Pulled in opposite ways by the twin forces of boldness and modesty, we have not known for sure in which way to direct our attack; and we have been thrown back, as a result, into postures of defensive self-criticism. Who are we humans to suppose that we can understand the nature of the entire cosmos, in the first place? And what kinds of knowledge or inference could cosmology legitimately aim at or rely on? How far can scientific observation, theorizing, and argument take us in a cosmological direction? And how far, therefore, is a theoretical account of the totality of things, rationally supported by human experience and knowable by human beings, *possible at all*?

II

The form of that final question—"How is cosmological knowledge *possible at all*?"—was, of course, made familiar by Immanuel Kant in the development of his mature "critical" or "transcendental" philosophy; and the case of Kant is specially illuminating for any discussion of cosmological method. Kant began as an enthusiastic cosmologist, but he made a drastic change of direction in mid-career; and the reasons for this change illustrate well the inescapable problems of substance and method which confront any honest and reflective thinker who struggles with problems about the structure, history, and destiny of the Universe for half a century or more.

During the 1740s and 1750s, Kant worked up a system of physical cosmology of a most daring and speculative kind; during the 1770s and 1780s, he switched to the new "transcendental" method of his three *Critiques*; and he concluded his work, at the turn of the nineteenth century, by returning to cosmology in a new spirit, in such late writings as the eschatological essay, *Die Ende Aller Dinger*. His first published essay (1747) was devoted to a central problem of physics which had vexed philosophers ever since the time of Descartes; namely, how one should properly define and measure such physical magnitudes as *motion, force*, or *energy*. (Specifically, he discussed the quantity that he called *lebendiges Kraft*—a phrase that we can best translate today as "actual energy of motion"—as contrasted, in good Aristotelian terms, with "potential energy.") But this was at most a preface to his first major book, the *Allgemeine Naturgeschichte und Theorie des Himmels* of 1755. This *Universal History of Nature* took as its foundation the central ideas of Newton's mechanical world picture, and used some more recent astronomical discoveries by Maupertuis and others as the material for a complete cosmogony (or historical cosmology) presented in purely physical terms.

From an initial condition of the cosmos, in which matter was distributed across unbounded space atom by atom in a random manner, Kant demonstrated how the universal gravitational force could have initiated an aggregation of massive material bodies, each of which would then become a local center of attractive force. These massive bodies would then, in turn, have congregated into whirling galaxies, including both our own local galaxy (the so-called Milky Way) and also Maupertuis's newly discovered *nebulae*; from time to time during astronomical history, gravitational collapse would have resulted in the violent local implosion of matter, followed by the explosive scattering of the atoms involved throughout the neighboring regions of space, as in a supernova; and so on, and so on. Putting forward a wealth of such imaginative hypotheses, Kant anticipated in his book many of the cosmological ideas that were to be explored by subsequent theoretical astronomers over the next two hundred years, and in this way he won himself an enduring place in the history of astronomy alongside Laplace and other major astronomical theorists of the late eighteenth century.

Kant had, of course, been able to develop this comprehensive account of the births, lives, and deaths of stars and galaxies only by extrapolating the results of Newton's work to the absolute limits of their validity, and even (perhaps) beyond. In this respect, he knew at first hand what it means *an die Grenze der Vernunft anzurennen*—that is, to run up against, and to overreach, the inherent boundaries of rational speculation—but his initial confidence in the necessary correctness of Newton's universal laws of physical nature shielded him against self-critical doubts. For it seemed as though Newton had succeeded, not only in recognizing the "empirical regularities" underlying all the phenomena of physics but also in expounding them in exact mathematical terms, and so had enabled us to draw "necessary inferences" about them. And, if Newton had really brought off this double success, that justified us in applauding his powers of insight and demonstration all the more.

From the 1760s on, however, Kant began to wonder whether Newton's double achievement could really have done all that he had originally assumed. Reading David Hume (he tells us) prompted a series of corrosive doubts which woke him from "dogmatic slumbers." How could a truly empirical, contingent, or synthetic natural science lend itself to a purely formal, mathematical, or necessary exposition? If Newton's physical system simply presented straightforward empirical discoveries about the world of facts, was it not incongruent to set out those discoveries— as Newton did—in necessary (or "apodeictic") mathematical arguments? If Hume were right, indeed, it would be especially hazardous to use Newton's theory as the basis for a comprehensive history of the entire universe, as Kant himself had done in the 1750s. Having digested the apparent consequences of Hume's questions, Kant recoiled from his earlier naive reliance on Newton's work—whether interpreted in an empiricist or a rationalist spirit—and began his painful reconstruction of philosophy on the new, "critical" basis. Only by recognizing and demonstrating the "transcendental preconditions" on which the very possibility of a mathematical science of natural phenomena depended (he now argued) could one bring to light, and learn to respect, the limits inherent in any speculative philosophy of nature.

Two seeming "necessities" of rational thought impelled Kant to change the direction of his scientific and philosophical inquiries in this way. In two fields of experience particularly—symbolized by "the starry heavens above" and "the moral law within"—he believed that the rational thinker encounters methods of thought whose "necessity" is seemingly inescapable. On the one hand, therefore, he set out to analyze the apparently "categorical" demands of morality; for example, the requirement that we seek to treat all other rational agents as ends in themselves and never as means only. On the other hand, the possibility of a rational natural philosophy posed an equally crucial problem; namely, to explain how the constructive activity of scientific reason could develop a mathematical theory of motion which was as necessary and rigorous as the geometry of Euclid, and which nonetheless succeeded in accounting for the actual, observed movements of the principal heavenly bodies. Kant's elliptical reference to "the starry heavens above" was in fact an allusion to the success of Newton's physical astronomy—a success that had been dreamed of by Plato long ago, called for once again in the 1620s in the philosophical manifestos of René Descartes, but substantially realized only in 1687, with Isaac Newton's *Mathematical Principles of Natural Philosophy*.

Kant's *Critique of Pure Reason* of 1781 therefore included, among other things, an explicit discussion of the conceptual limitations to which all our cosmological speculations are subject. Spatial and temporal notions, for example, play indispensable parts in the intellectual ordering of our experience, in technical physics and in everyday life equally; but that fact, by itself, does not give us an unrestricted license to elevate Space and Time into reified entities—still less, to treat space-as-a-whole and time-as-a-whole, either as straightforward objects of experience or as topics for scientific hypothesis. On the contrary, incautious extrapolation of our familiar spatio-temporal notions (Kant warned) can lead us into fallacies of a special kind, which he called *paralogisms*. There is (he argued) something intrinsically paradoxical about such phrases as "the Boundary of Space" and "the Beginning of Time." We know how to measure the beginning of any single physical process taking place—as we say—in time; but against what larger measure can we meaningfully speak of "Time itself" as beginning? For any argument designed to establish that Time itself has a beginning, or the totality of Space a boundary, a counterargument can be developed to establish that Time, in itself, is endless and Space unbounded.

not!

Kant's critical philosophy thus focused attention on those arbitrary elements that are involved in dividing up the physical world *exactly* into spatial and temporal *parts*, according to our own chosen definitions, scales, and methods of representation. But other equally arbitrary subdivisions can also mislead us, as we set about constructing a comprehensive and integrated account of the Universe as a Whole. These subdivisions separate out *aspects* of the natural world, rather than parts: distinguishing between the varied scientific questions that arise for the purposes, and from the standpoints, of different scientific disciplines. Approaching the world of nature from the standpoint of electromagnetic theory, for instance, we have occasion to ask questions about *fields* and *charges*, *currents* and *radiation*; approaching it as classical chemists, we have occasion to discuss *atoms* and *molecules*, *elements* and *compounds*; approaching it from the standpoint of evolutionary biology, we have occasion to ask questions about *genotypes*, *mutations*, *selection pressures*, and the like; but the disciplinary organization of scientific work deprives us of any standpoint from which to ask fully comprehensive questions, transcending the particular standpoint of any single discipline. Even today, indeed, natural scientists still tend to assume that the only truth about the Totality of Nature is the totality of the established truths about the different disciplinary aspects of nature, or sets of natural phenomena, taken one at a time. Add together the existing discoveries about evolutionary, electromagnetic, and other "aspects" of nature taken separately, and what more is there to be said about the universe?

If we are to establish the full "preconditions" for a rational cosmology, therefore, we have first to face one final "transcendental" question:

> In what respects, and on what conditions, can anything be said about the natural world *in its entirety* which is not dependent on our ability to subdivide natural phenomena into separate aspects, along disciplinary lines, and discover truths about those aspects one at a time?

& what about
the concept
of entity?

Is "the Universe as a Whole," so considered, a legitimate topic of rational investigation and speculation, or even a legitimate subject of human language at all? Or have we deprived ourselves of that possibility in the act of setting up, for analytical purposes, the very subdivisions of nature which define the individual disciplines of science? Are there, after all, things to be discovered about the entire universe, or about the place of humanity within that entirety, whose very generality allows them to transcend the fragmented insights of electromagnetic theory, cell biology, neurophysiology, and the rest? Or does the stern injunction admit of no appeal?— "What Science hath put asunder let no mere Theologian seek to join together again!"

III

What kind of things, then, can we learn from the results of the natural sciences which are of cosmological significance today? That question is still as problematic as it ever was, and the fallacious shortcuts by which we are tempted to circumvent those problems remain as enticing as ever. For in this respect (it seems) our tasks can be simplified in either of two different ways. Let us look at these in turn.

On the one hand, we are tempted to take the theories and concepts of the sciences entirely at their face value in a perfectly naive manner: extrapolating them without hesitation beyond their original scope or range, as far as our cosmological purposes require, without regard to Kant's warnings against overrunning the inherent limits of rational investigation. This first shortcut has the merit of allowing our cosmological imaginations free rein; but this advantage carries with it corresponding risks, in particular the risk of overinterpreting the scientific concepts concerned. Struck by the intellectual power of atomistic explanations in physics and chemistry, for instance, we may too readily assume that the Universe as a Whole is therefore constructed entirely of atoms; whereas the first task is to inquire how far the atomistic concepts that are so fruitful in physics and chemistry can safely be extrapolated and reapplied, not just to the phenomena of gas theory and chemical combination but to the totality of things.

This first kind of shortcut has a long and venerable history. Quite apart from the traditional cosmologies of the Middle East, which dated back long before the time when Greeks of Ionia and Italy inaugurated formal philosophy, we might regard the pre-Socratics—Thales and Empedocles, Pythagoras and Anaxagoras—as cosmologists whose views of nature extrapolated conceptions whose primary application was to particular kinds of natural phenomena, and applied them to the entirety of nature. For, as Aristotle himself remarked, in the first century or two of philosophizing, the line separating mythology from philosophy was a very thin one.

With the renewal of scientific speculation from the sixteenth century A.D. on, cosmological ambitions revived alongside various explanatory ones. In the 1690s, for instance, Isaac Newton's pupil William Whiston produced a speculative account of the history of the astronomical universe, based on Newton's theories, which anticipated some of the most striking features of Immanuel Velikovsky's recent theories—for example, explaining the biblical Flood as the product of gravitational forces associated with a planetary "near miss"—and another of Newton's contemporaries, Thomas Burnet, did the same for geology in his *Sacred History of the Earth*. (In this respect, Kant's early cosmological writings were less original in their scope than they were in the rigorously scientific quality of their arguments.)

The self-critical objections to unbridled cosmologizing which Kant raised in his *Critiques* were not, however, the end of the matter. On the contrary, the nineteenth and twentieth centuries have generated their own scientifically based cosmologies, and the precise nature of the inherent limits on cosmological speculation has remained as much in dispute as ever. During the nineteenth century, atomism, historical geology, and evolution theory took over the leading place in cosmology from astronomy; and the public dispute about Darwinism, for instance, turned as much around the larger implications of natural selection for our overall view of nature as it did around the explanatory power of Darwin's analysis for the specific problems of biological analysis. So, toward the end of the century, we find several striking but inconclusive attempts to generalize the results of contemporary science and raise them to the cosmological level: Friedrich Engels's *Dialectics of Nature* is one example, Ernst Haeckel's enormously popular *Riddle of the Universe* is another. (Several twentieth-century examples are the topics of the essays collected in Part Two, below.)

On the other hand, a fear of falling into ''scientism'' may lead us to draw back, and to dismiss the natural sciences as a source of material for cosmology. Science (we may be tempted to argue) provides us only with sets of intellectual tools having well-defined but narrow uses, and these pragmatic instruments are of no relevance to the more speculative purposes of cosmology. This skeptical mode of arguing, too, is an ancient and venerable one. Even if we do not count Socrates as a true skeptic about cosmology, we can find the argument relied on by Claudius Ptolemy, as early as A.D. 150. Ptolemy well understood the intellectual shortcomings of his great astronomical synthesis, the *Megiste Syntaxis* or *Almagest*. He had not been able to present a unitary, systematic, or even wholly consistent account of the movements of the different planetary bodies. Instead, he had collected a range of largely unrelated algorithms, or computational procedures, for calculating different aspects of those motions separately; and the physics of the underlying processes he had left a mystery. But after all, Ptolemy commented, human observers and thinkers are terrestrial beings, and so have no access to *to theion* (the ''celestial'' or ''divine'' part of the cosmos); they are therefore in no position to speculate about the celestial regions with any hope of arriving at either Truth or Reality.

A similar argument was put forward by Andreas Osiander in the mid-sixteenth century, when he sent Nicolaus Copernicus's *De Revolutionibus* to the printer with a covering note of his own which denied—in a very un-Copernican spirit—that the book claimed any "truth" or "reality" for its explanations. All that Copernicus intended to do (Osiander declared) was to develop improved algorithms for convenience of computation, without giving them any realistic interpretation in physical or cosmological terms. When we recall how Galileo was subsequently harassed for doing little more than Copernicus had done, we may recognize the diplomatic purposes of Osiander's note. Yet, in retrospect, it is clear enough that Osiander was excessively cautious, and that his way of presenting Copernicus's arguments needlessly blunted their impact and weakened their force.

Much nearer to our own time, in the early twentieth century, one finds the same kind of position adopted by Pierre Duhem, the French thermodynamicist and historian of science. Throughout all his intellectual work—as he made clear in his essay *Physique d'un Croyant* (*The Physics of a Believer*)—Duhem was concerned to prevent his professional commitments as a physicist from coming into any direct conflict with his personal commitments as a Roman Catholic. The believing Catholic in him reserved to religion any right to speak about Reality, and the physical scientist in him therefore made severely limited intellectual claims for the results of the natural sciences. Scientific hypotheses and theories (he argued) are merely intellectual instruments that human thinkers construct and use to relate together their limited observations of nature. The only purpose of those hypotheses and theories (to use the classical Greek phrase) is to "save the appearances," and it is a mistake to assume that they tell us anything about "reality"—let alone about the *entirety* of reality. Extrapolating the restricted concepts and hypotheses of any science, from the self-limited domain of phenomena proper to the discipline onto a universal or cosmological scale, will therefore be to take an unjustified leap from "appearances" to "realities," and so involve an illegitimate inference.

However, neither of these two shortcuts—either taking scientific results entirely at their face value or else ignoring them entirely—is fully acceptable today. Our cosmological ideas about the universe, and about the place of humanity within that universe, cannot simply ignore Science; instead they must surely be framed in terms that make the best possible sense when viewed in the light of our scientific results, without overextending the scientific concepts in question. We cannot afford to embrace the results of all the specialized scientific disciplines naively and uncritically; but neither can we dismiss them as completely irrelevant, in principle, to the whole cosmological project. Rather, we need to look for a middle way: considering with more discrimination just which scientific concepts and hypotheses are directly relevant to cosmological issues, and with what qualifications they can be given this wider application.

The essays collected in this volume, accordingly, map the changing relations between science and theology in the field of cosmology since the 1950s. They are concerned above all with this task, of finding a "middle way" between the credulity of a Haeckel and the skepticism of a Duhem—a way ahead which requires us neither to deny completely nor to exaggerate incautiously the cosmological implications of twentieth-century scientific thought. These essays neither identify the concerns of cosmology with those of science nor do they try to separate them entirely. Instead they attempt, in a preliminary way, to answer the transcendental question, "On what preconditions is a science-based cosmology *possible at all?*"

In this respect, indeed, they are the products of both a professional and a personal quest; and the arguments they present, from the early 1950s to the late 1970s, are historical cross-sections both of the collective history of cosmological ideas, and also of the author's own intellectual development. The earliest essay, reprinted as Part One, tilts somewhat more toward skepticism than toward credulity; while the most recent one, Part Three of this volume, proposes a larger and more open role for scientific ideas in the future development of our cosmology.

Professionally speaking, this shift of emphasis reflects a broader shift in our intellectual and cultural attitudes which began in the mid-1960s and has not yet exhausted itself: a shift from a single-minded preoccupation with intellectual purity and abstract theory to a more complex concern for human relevance and concrete historicity. Thirty years ago, the separateness of different intellectual disciplines was an unquestioned axiom of intellectual procedure, and the obstacles to thinking of the natural world in other than strict, disciplinary terms were still very substantial. Around

1950, accordingly, caution was the order of the day. Within fundamental physics, for instance, David Bohm's critical questions about the finality of the quantum-mechanical world view were widely dismissed as being not merely heretical but downright perverse, while in cosmology it appeared more important to control one's scientific enthusiasm than to give free rein to one's speculative imagination.

By the mid-1970s, the cultural atmosphere had become more relaxed. Even within the natural sciences proper, a shift from narrowly disciplinary preoccupations to more interdisciplinary issues had at least made it possible to reopen, in a serious spirit, questions about the cosmological significance of the scientific world picture. Within fundamental physics, John Wheeler was taking on the role of standard-bearer for a participatory view of our place in the natural world, as Laplace had done for the onlooker's view much earlier on; and by doing so he was helping to lower the barriers separating scientific cosmology from natural religion. So the disciplinary specialization of the natural sciences can no longer intimidate us into setting religious cosmology aside as "unscientific." Instead scientists, philosophers, and theologians alike can embark, with sober courage and a confidence that by the standards of 1950 would have struck us as naively euphoric, on the shared task of constructing a conception of "the overall scheme of things" which will stand up to criticism from all three directions.

As early as 1944, I recall sending a letter to the scientific weekly *Nature* which the editor had the good sense to reject. This letter bore the heading, "Ethics, Evolution and Thermodynamics," and it comprised the outline of a comprehensive cosmology that attempted to relate the historical limits on human choice and destiny to the scientific conceptions of "entropy increase" and "evolutionary progress." If that 1944 letter sketched my own personal *Allgemeine Naturgeschichte*, then eighteen months of study with Ludwig Wittgenstein immediately after the Second World War did for me what a reading of David Hume had done for Kant. By 1948, I was in full reaction against the intellectual simplicities of my earlier cosmological speculations, and wrote a brash dissection of Julian Huxley's "evolutionary ethics" under the title "World-Stuff and Nonsense."

Some of the arguments of that *Cambridge Journal* review survived in only somewhat less harsh form as part of a longer essay on "scientific mythology," which I published in 1957 and which is reprinted as Part One of the present collection. The reservations I had come to feel under the influence of Wittgenstein about the naive extrapolation of scientific concepts into nonscientific contexts still needed to be taken seriously, I thought; and, at that time, the changes within the natural sciences that would make me more optimistic about circumventing Wittgenstein's reservations had not yet occurred.

In Part Two of this collection, several essays on individual scientific cosmologists appear in chronological order, and readers will note a gradual change of tone. Through the 1960s, I remained in some doubt whether the cosmological theories that would be arrived at, if we relied on the ideas of the natural sciences to underpin our larger hopes and dreams, had any real intellectual value or made any real sense. In the case of Teilhard de Chardin, those doubts of mine are still active. His whole charming scheme of "noosphere," "omega point," and the rest, still strikes me as being a vehicle of wish fulfillment as much as of serious thought; and his grand vision of the biological world as engaged in an overall project of self-perfection owed more to the creative imagination of Lamarck and his philosophical successors in France—notably, Bergson—than it did to the serious work and analysis of those evolutionary biologists whose results Teilhard de Chardin appealed to as authoritative. (Not that Teilhard's personal extrapolation of evolutionary ideas onto the cosmological scale was an exclusively French heresy: Teilhard wrote with an inimitable French flair; but a similar mix of ideas is found in the Englishman Julian Huxley.)

Still, as John Dawson recognized when he fingered him to unearth the counterfeit jaw bone at Piltdown, Teilhard was too easy a target. Another more serious aspect of scientific thought in France is represented by Jacques Monod and François Jacob, whose cosmological books (1971 and 1974) come more closely to grips with the larger implications of Darwinism, and address issues of cosmology and methodology in terms that demand more careful scrutiny. Their writings illustrate in an interesting way the continuing influence of René Descartes's mechanistic view of the physical and biological worlds, which for many years made Darwin's own historical patterns of explanation uncongenial to French professional biologists—at least, until the decipherment of the "genetic code" carried in the nucleic acid macromolecules gave "evolution" a new and mechanistically intelligible material basis.

This emphasis on the peculiar Frenchness of Chardin, Monod, and Jacob is not meant frivolously. Even within the stricter traditions of scientific theory proper, there is some room for differences of national, local, and even personal style; so that we find Albert Einstein (for instance) manifesting a concern for the role of "symmetries" in physical nature, which to begin with was quite personal to him, and may even—as Gerald Holton has argued—have reflected his idiosyncratic habits of "visual" rather than "verbal" thinking. Within the larger and less strictly controlled speculations of the scientific cosmologists, however, these stylistic variations are par for the course. Just because such theories use the language and ideas of the sciences to express a more broadly synthetic vision of Humanity's Place in Nature, they depend that much more on the earlier backgrounds and presuppositions that the writers *bring to* this task. The reading of Arthur Koestler's overall scheme (or "holarchy") to which I finally came in 1979, at the third attempt, is likewise intended in all seriousness. The selective way in which Koestler picks and chooses between those scientific views that he can be happy with, and those that he feels bound to condemn, certainly rested (in my view) on demands that he was bringing to science from outside, rather than on the intrinsic character of the scientific ideas in question. The interpretive challenge, therefore, was to think up a reasonable hypothesis about the considerations that underlay, and gave force to, those external demands.

It was not until the late 1970s that I was ready to return to the cosmological writings of scientists in a fully sanguine and constructive spirit: the reviews of books by Carl Sagan and Gregory Bateson, reprinted here, benefited from that change. By that time, the harshest rigors of both positivist philosophy of science and its more dogmatic critics were fading into memory; and, in the more relaxed intellectual atmosphere so created, it was possible to reconsider a whole range of forbidden topics. Maybe, after all, the strict regimen of disciplinary science had won authority only on certain conditions, and at something of a price. If so, then maybe there was still a real chance of working outward from the natural sciences, and into the larger cosmological realm. But the first step was to clarify the conditions and limitations on the intellectual authority of the traditional scientific disciplines. When I started writing the 1979 Tate Willson Lectures for Southern Methodist University, with which this collection concludes, that was accordingly the question from which my argument began.

IV

Cosmology may be a frustrating area of intellectual inquiry, but it need not remain forever a source of disappointment or disillusion. If we succeed in traveling any distance along the new roads that are now opening up to us, cosmological discussion will be a little less frustrating, and will bring us results somewhat more proportionate to the intrinsic fascination of the subject. For theologians, it will offer a way of moving down from the high places of transcendental metaphysics, in which they have too often taken refuge during the last two hundred years, and allow them to take their place once again in the world of concrete experience, within which alone their arguments can either strike a real spark in the minds of their hearers or carry real conviction in their hearts. For scientists, it will open up a field in which they can once again take up, on equal terms, that dialogue with the philosophers and theologians which Descartes and Leibniz, Newton, and Priestley carried on unhesitatingly in the days before the disciplinary fragmentation of the scientific debate. For philosophers, too, it will provide a forum in which they can enlarge their agendas, combining the critical techniques of the analytical and phenomenological traditions in the service of a larger and more constructive enterprise. After all, it will be no insult to a Heidegger or a Wittgenstein if we now seek to move beyond their critical insights and take up again the larger cosmological issues that formed a traditional part of the philosophical agenda from the time of Aristotle on.

True, if we do so, we shall have to frame our questions in terms that are adapted to the historical conception of nature current in our own day, just as Aristotle did for the ahistorical conception of nature current in classical antiquity. (Too much has happened during the last hundred years, in any event, for us to leave Friedrich Engels's *Dialectics of Nature* in undisputed command of this field.) By the early 1980s, however, we have done little more than reestablish the bare preconditions for the development of such a new cosmology. We can see that the arguments that for so long prevented any legitimate union of ''natural science'' with ''natural religion'' were arguments whose validity was at best conditional and temporary, and we can recognize what fruits such a renewed union promises on a variety of different levels from pure theory to practical politics. Still, it is one thing to reestablish the preconditions for such a unified world view. But it is quite another thing to develop—and to find effective ways of expressing—the true sense of cosmic proportion that will be called for in any satisfactory new account of ''the overall scheme of things.'' That will be the task for other writers and thinkers and for much larger books than this one.

PART ONE

SCIENTIFIC MYTHOLOGY

Originally published as "Contemporary Scientific Mythology," in *Metaphysical Beliefs*, ed. Alasdair MacIntyre (London: SCM Press, 1957).

Scientific Theories and Scientific Myths

If we go into an eighteenth-century library, we may be surprised at the number of theological works it contains. Baxter's *Reasons*, Ogden's *Articles*, Warburton's *Divine Legation*: there they stand, and with them the sermons, row on row of them, solid, calf-bound, imposing; yet somehow (we feel) period pieces, as foreign to us in our day as the wigs and top-boots in a Hogarth print. For a member of Dr Johnson's Literary Club, it was as important to be *au fait* with Ogden or Warburton as it was to be ready with an apt quotation from Pope or Horace. Anyone who has read his Boswell knows how often, when gossip was exhausted, conversation in Johnson's circle turned to ethics, philosophy or theology; for these were subjects in which any educated man felt an obligation to be interested.

We in the twentieth century, however, feel different obligations. It is science we like to be up-to-date in, Freud and Hoyle we choose to know about. We are interested less in the doctrine of the First Cause than in physical cosmology, while the Ten Commandments and the nature of the moral sense seem tepid to us when set alongside the theory of the super-ego, *Autres temps autres mœurs*: the emphasis in polite conversation has shifted. If we are puzzled by the shelves of collected sermons in our ancestors' libraries, that is because we forget how far scientific and aesthetic questions have replaced moral and theological ones as the staple of dinner-table-talk; and how far the popular scientist has won over the audience of the popular preacher.

At first sight, this appears a remarkable change, and certainly, so far as prestige is concerned, science has made great advances at the expense of philosophy and theology: that much is a commonplace. But is the change as great as it seems? Are people really no longer interested in all those serious topics which preoccupied their ancestors, finding themselves absorbed instead in some quite different set of problems? Or are the same old cargoes being carried (so to speak) in fresh bottoms, under a new flag? What answer we give to that question depends on this: how far the problems the man-in-the-street expects the scientist to solve for him are ones about which a scientist is specially qualified to speak. So before we are too impressed by the change it is worth asking whether, when we turn to works of popular science, the questions we are interested in are always genuinely scientific ones. I think this is only partly so, and in what follows I shall try to show why. Often enough, we tend to ask too much of science, and to read into the things the scientist tells us implications that are not there—which in the nature of the case cannot be there; drawing from scraps of information about, for instance, physics, conclusions which no amount of physics could by itself establish. Sometimes our questions are clearly the same as those that the eighteenth-century theologians tackled: a discussion about free-will is none the less about free-will for bringing in Heisenberg's 'uncertainty relation'. But more often we are unaware of what we are doing, and turn to the scientist as to an expert, an authority, even when he is entitled to no more than a private opinion.

Quite a lot of popular science books encourage us in this, and present these opinions as the latest results of scientific research. Their authors do not confine themselves to explaining some scientific investigation, some novel theory or discovery about phenomena which had previously not been understood. They go on to do something more, something different, something which can hardly be called science at all. As a result there has grown up a sort of scientific harlequinade in the shape of an independent body of ideas which play a considerable part in the layman's picture of science, but in science proper none at all. The Running-Down Universe, Evolution with a capital E: these are two examples which (I shall try to show) are not so much scientific discoveries as scientific myths.

'Scientific Myths': the very phrase is apt to sound a little paradoxical. For we like to think of myths as a thing of the past. We pride ourselves that they have been killed, and killed, furthermore, by science. Atlas, Ceres, Wotan, Poseidon . . . *nous n'avons pas besoin de ces hypothèses*. These names are for us the last relics of an outmoded system of thought, which attempted to explain in one way—by personification—things which we can now explain much better in another. The stability of the earth, the fertility of the soil, the ever-varying behaviour of the sea, these are all things we understand well enough nowadays without the need to bring in giants and goddesses.

This view of myths is, however, a shallow one. The attempt to explain natural phenomena by personification may be dead, or moribund. But many of the motives which produced the myths of the Greek and Northern peoples remain active in us still. In consequence it is not enough to regard the old stories only as half-baked science. They were that, no doubt, among other things. When people used to talk about Zeus or Wotan as the thunder-maker, they certainly thought that in these terms the occurrence of thunderstorms could be explained, so to this extent the notion of Zeus played for them the part which the notion of atmospheric electricity does for us. Variations in natural phenomena, failure of the harvest or turbulence of the sea, were likewise to be understood in anthropomorphic terms, as the moods of divine agents, Ceres or Poseidon. But there was always something more to these myths. Zeus was not only the thunder-maker, he was also the Father of Men; and as such he played a very different role. For mere disinterested curiosity over unexplained phenomena would never have led people to talk of a 'divine father', whether in Heaven or on Olympus: that has never been a purely scientific conception. So, though with the progress of science the motions of the sea and the stars and the growth of the corn have ceased to be for us the work of hidden hands, nevertheless some of the motives for myth-making are with us to-day as much as ever they were. Myths are with us, too. Our difficulty is, to know in which direction to look for the myths of the twentieth century, to recognize and unravel the nonscientific motives behind them, and to see these motives at work.

If we do think ourselves myth-free, when we are not, that is (I am suggesting) largely because the material from which we construct our myths is taken from the sciences themselves. The situation is the one we meet in those trickiest of crime stories, in which the detective himself turns out to have done the deed: he is the last man we suspect. There are of course other reasons too why we find it hard to recognize our own myths. To begin with, they are hard to spot, as our own fallacies are hard to spot, just because they are our own: fallacies, we are tempted to think, are the faults in *other* people's arguments, and myths the queer ideas people *used* to have about the universe. Again, we are inclined to suppose that myths must necessarily be anthropomorphic, and that personification is the unique road to myth. But this assumption is baseless: the myths of the twentieth century, as we shall see, are not so much anthropomorphic as mechanomorphic. And why, after all, should not the purposes of myth be served as effectively by picturing the world in terms of mythical machines as by invoking mythical personages? Still, in the main, it is because our contemporary myths are scientific ones that we fail to acknowledge them as being myths at all. The old picture of the world has been swept away; Poseidon and Wotan have suffered death by ridicule; and people not unnaturally look to the scientist for a substitute.

Therein lies the misunderstanding, for only in part were the ancient myths half-baked science, and only in part was their role an explanatory one. So far as this was so, we can reasonably look on the natural sciences as their descendants; but only so far. The other non-scientific motives behind them remain, and the sciences are not obliged to cater to these. The notion of atmospheric electricity, for example, was introduced to account in a scientific way for lightning and thunder, and to that extent displaced Zeus as the thunder-maker, but it was never intended to take over Zeus' role as the 'divine father' as well. Rather, the two roles have been separated, so that thunderstorms are no longer regarded in the old way, as a topic for theology.

It is not enough, however, to suspect that there may be such 'scientific myths': we must also know how to recognize them when we come across them. How are we to do this? Partly, I have argued, by seeing what sorts of questions they are used to answer: if a conception, however scientific its birth or ancestry, is used in practice only as a way of dealing with non-scientific questions—whether ethical, philosophical or theological—then it is no longer following the trade of its forefathers, and has ceased itself to be a scientific term at all. Again, there are some terms of irreproachably scientific origin which begin after a time to live double lives: as well as their primary, scientific *métier* they acquire part-time jobs of another kind. If we find evidence of such duplicity, our suspicions will be confirmed.

This is a clue, but it is one which immediately raises further questions. How is it, for instance, that such a double life is possible? Scientists take so much care in defining their terms that serious ambiguities cannot, surely, remain: if the meaning of their terms were not clear, one would expect this to have its effect on their work—to lead, that is, to trouble within science itself. And in any case, if a scientist has been true to his declared method, and has introduced into his theories only those terms which he absolutely requires in order to explain the phenomena he has been studying, what room is there for equivocation?

The answer to this question is a double one, partly historical, partly logical. The ideal of a science which contains nothing but what is forced on us by the phenomena we are studying is only an ideal: it is not, and never will be, an accomplished fact. As a matter of method, no doubt, scientists do develop and modify their theories and conceptions in just such ways as will (so far as they can see) best accommodate the phenomena; nor are they prepared to allow outside considerations to obstruct such developments as the phenomena require. But the theories they subscribe to as a result, whether in the sixth century B.C. or in the fifteenth or twentieth A.D., fall short of the purist's ideal for two reasons. To begin with, their historical origins are against them. Anyone can see the points of resemblance between the cosmology of Plato's *Timaeus* and the Near Eastern myths which it was intended to displace; and, though many of these residual elements of myth have since dropped out of our science, it is imprudent to point the finger of scorn at Plato (as Sarton does) until we have inquired whether this elimination has been completed. It is wiser to recognize that, as our scientific ideas develop, there will always be a tug-of-war between tradition and method: a scientist's *methods* may be completely empirical, yet his investigations will have no direction without the guidance of a pre-existing body of ideas, some of which may turn out under scrutiny to be survivals from surprisingly far back.

This factor may, as the centuries pass, be of less and less importance, but the other is of permanent relevance. However much the sciences may eventually outgrow their historical swaddling-bands, there must always be something more to the framework of ideas which constitutes a theory than the bare recapitulation of the phenomena it is used to explain. The structure of a scientific theory may be built up entirely from the bricks of observation, but the exact position the bricks occupy depends on the layout of the scientist's conceptual scaffolding; and this element of scaffolding, which the scientist introduces himself, is always open to misinterpretation.

Neither of these factors is one which need affect the scientific value and validity of a theory. If a term like 'evolution' comes to be used ambiguously—having both a pure biological use and an extended, philosophical or mythological use—this ambiguity is not one which will necessarily show up in a strictly biological argument. The aspects of the notion which are put to mythological use may not be ones that bear either way on any biological questions; and so long as they do not do so, the notion will preserve all its power within biology. Even to speak of ambiguity in this context may therefore be too strong. What we have rather is a choice between two interpretations of a term, a narrow one and a wide one: a narrow one, whose use and justification lie wholly within the natural sciences, and a wider, extended one, whose justification and use both lie in part elsewhere.

With this point in mind, we can clear out of the way one elementary misunderstanding. When I go on to argue that some familiar notion—The Running-Down Universe, for instance, or Evolution regarded as 'the Cosmic Process'—is a scientific myth, I shall not be making a point which raises questions of a straightforward scientific kind. In particular, I shall not be casting any shadow of doubt either on the laws of thermodynamics, or on the doctrine that species have developed by variation and natural selection. There are, no doubt, plenty of people who still reject Darwin's theory even as biology. I am not one of them: in its essentials it seems to me among the finest and most firmly established products of biological thought. The claim that Evolution is sometimes treated as a myth must not, then, be misunderstood: it is quite distinct from any possible claim that, as a scientific theory, there is something dubious or unsound or even speculative about the Darwinian view of the origin of species.

People do, it is true, sometimes say 'So-and-So is a myth', meaning only that the belief is untrue or unsound; and this might be said of Darwin's theory by an anti-evolutionist, as a contemptuous way of dismissing it. But do not let us fall victims to this sort of loose expression. To use the word myth only as a term of abuse is to rob ourselves of a useful distinction. Not all outdated scientific concepts were myths, nor vice versa—caloric, for instance, had no mythological significance. So, if we talk about scientific myths, let us do so strictly; in order to raise not scientific issues but logical ones. Granted that the theory of evolution or the laws of thermodynamics are all that a scientist can ask; granted that their position within biology and physics is as firmly established as it could be; if this may be allowed, just how much is accomplished? What sorts of conclusions are forced on us by our acceptance of these theories, and on which do they have no direct bearing? These are the questions we must ask. If we find that the theories are regularly invoked in support of conclusions of a kind to which, as scientific theories, they have no relevance; further, if these conclusions are of a sort with which mythologies have from the earliest times been concerned; then we can say with some justice, not that the theories themselves are 'only myths', but rather that on these occasions their conceptions are being inflated into Scientific Myths.

Once again, then, how are we to recognize when a scientific term is being pressed into service of a non-scientific kind? The chief point to look out for is the following. When a technical term is introduced into a science, or an everyday word like force or energy is given a fresh, scientific application, it has a clearly defined place in a theory—a theory whose task it is to explain some limited range of phenomena. What gives the term a meaning for science is the part it plays in these explanations. One can think of such a term as a piece in a jig-saw puzzle; and, like such a piece, it loses most of its significance as soon as we try to make anything of it out of context.

We can take the notion of universal gravitation (gravity, for short) as an example. When Newton introduced this idea, his purpose was a limited and tangible one: namely, to account for the motions of the planets, the comets and the moon in terms of the same laws of motion as held for terrestrial bodies. And when one says 'account for', this means (as he himself took care to emphasize)[1] account in a mathematical way. For Newton's purposes, the term 'gravity' acquired its meaning with the introduction of the inverse-square law; and this in its turn earned a place in physical theory because it could be used to work out how, in this or that situation, celestial or terrestrial bodies can be expected to move. As a piece of planetary dynamics, Newton's theory needed no other justification. He saw that in due course the theory might be amplified to deal with other phenomena, and the mode of action of gravity might thereby be discovered; but, he insisted, we must not jump to conclusions; and in any case his notion of 'gravity' should not be taken as having any implications outside dynamical theory.

It is to some such modest but solid job that all scientific terms are put, 'evolution', 'entropy' and so on, quite as much as 'gravity', and it is vital for the progress of science that their meaning should be limited in this way. It is just because the terms of the sciences are so well defined, and defined in a way which is closely tied down to the phenomena, that questions in science can be settled: only because this is so can scientists hope to answer definitely the questions that arise for them, by looking to see whether things actually happen in nature in the manner the theory suggests—in this way, they can usually come to agree upon one answer and reject the alternatives.

If this is forgotten, difficulties are created. Suppose we extend a carefully defined scientific term beyond the range of its theory, and use it in more ambitious but less tangible speculations, then there will be snags at once. For whereas before this was done one could check the soundness of one's speculations against the facts, now things will be different: there will be no way of checking what is said by experiment or observation, and so, scientifically speaking, nothing to choose between one possible answer and another. And if

1. Cf.: Newton, *Mathematical Principles of Natural Philosophy*, ed. F. Cajori (1934), pp. 550–1: "The System of the World," §2.

this is so, if when a dispute arises there are now no conceivable observations to be made by which we can decide between the disputants, then there can be no question of either side in the dispute claiming for his doctrine the support of the theory concerned. The theory will be neutral between all such views.

Newton realized this also. It was not that he had any objection to wider speculations: as we know now, he spent a surprisingly large part of his time on natural theology, the interpretation of biblical prophecies and other non-scientific problems. But in these speculations he did not keep appealing to 'gravity'. The term had a clear meaning in dynamics, and it could play a part in theology only if it were given a radically different sense. When he came to expound his theory of gravitation, therefore, he put aside all wider questions, and the only rival views he bothered to consider were other genuinely physical ones—for instance, those of Descartes and his followers.

In Descartes's picture of the solar system one must think of the sun as surrounded by a vortex, and of the planets as carried round about it like floating chips: the idea of gravitational attraction played no part in the account at all. To this theory, Newton's reply was simple. It is not enough, he argued, for a theory to provide a vivid picture of the solar system: one must work out the mathematical consequences of the view in detail. If this is done for the vortex theory, you cannot, short of the most implausible and groundless assumptions, make it fit the facts. In the first place, to talk of a vortex at all suggests that the space between the planets is filled with some kind of celestial bath-water, whose motions carry the planets round with it. But there is no independent evidence at all for supposing the existence of this fluid: indeed there are several reasons for rejecting the supposition—such as the fact that comets travel right across the solar system without showing any sign either of resistance from the fluid or of the effects of the vortex on their line of travel, and the fact that the satellites of each planet, however far it is from the sun, travel round it in the same manner. Further, to make the vortex theory work quantitatively, one must assume not merely the existence of this wholly impalpable fluid: one must imagine it endowed with physical properties (already, alas, indetectable) which vary greatly from point to point in space. A theory expressed in such terms as these could be of little use to science. Newton's own theory, by contrast, would account for all the observed motions of the comets, the planets and their satellites exactly, and without such a mass of arbitrary assumptions.[2]

No wonder Newton felt entitled to be satisfied with his theory. Yet it was assailed at once, from several directions. The followers of Descartes, of course, objected to the theory as physics; but others found wider reasons for attacking it. Leibniz, for instance, accused the doctrine of universal gravitation of being repugnant to common sense: to speak of the heavenly bodies as gravitating towards one another was, he said, 'a strange fiction'.[3] He further agreed with Berkeley in finding the implications of Newton's views impious, if not actually atheistical, for reasons which we do not now find it easy to accept.

These wider criticisms distressed Newton, but he did not spend much time answering them. Of course what he spoke of as gravity was an extension of the everyday, terrestrial notion, and must be understood as such. Leibniz might want to stand by the old sense of the term—the one enshrined in seventeenth-century 'common sense'; but if one looked at the uses to which Newton put his extended notion, one would see how the extension could be justified.

Before his time, the notion of gravity had an application only to bodies on the earth—pick up a chair and it feels heavy (*gravis*), let it go and it falls (gravitates) to the ground. The heavenly bodies, by contrast, were quite unlike chairs. They moved in their stable orbits round the sun or kept their places in the more distant firmament— the notion of gravitation manifestly had no relevance to their behaviour. The hypothesis that the planets were *massive* was no doubt an intelligible one; that they were *heavy* would have been a less intelligible suggestion; and to talk of heavenly bodies falling or gravitating would have called to mind only the falling stars which appeared in spring and autumn to drop through the night sky, or the thunderbolts which from time to time would strike the earth and awe the superstitious.

2. Cf.: *op cit.*, pp. 385−96; Bk. II, sect. ix.
3. Leibniz-Clarke correspondence: Leibniz's fifth letter, §35.

Newton's theory changed all that. The regular motion of the planets round the sun, which his predecessors had so carefully described—this too, he declared, was an effect of 'gravity' and a case of 'gravitation', just as much as the weight of a chair or its fall to the ground when released. The view might seem paradoxical to some: the stars and planets have no visible means of support, yet they do not, like terrestrial bodies, fall to the ground for lack of it. But the view has a point, and a *scientific* point at that. One can represent the motions of the planets round the sun with a degree of accuracy exceeding anything detectable by observation in New-ton's time by regarding them as freely moving bodies, acted on only by his 'inverse square force': exactly the same force can be appealed to in explanation of those terrestrial phenomena which alone had hitherto been called gravitational. This was all that was in question in calling the motion of the planets 'gravitational', or an effect of 'gravity'.

As for the question, whether the new theory was atheistical or not, even to ask this was to read things into the theory. What Newton had been doing was a piece of physics, as a result of which he had been able to explain in a mathematical way how the planets moved. The solar system could, he thought, be none the less wonderful—none the less a tribute to the foresight of the Almighty—for our having gone thus far towards understanding it. Indeed he himself was inclined to think the opposite. To have shown that one set of mathematical principles underlay so many varied dynamical phenomena should (as he put it) 'work with considering men for the belief of a Deity', so he could see nothing impious in the theory.[4] In any case, when it came to natural theology, what told in the balance was not the details of the theory. The precise form of his law of gravity could not therefore be relevant to any theological issue: the success of an inverse-cube law would have been no less impressive than that of his own theory. All that was at issue, for theology, was the rationality of the universe, and this was something which any successful and comprehensive theory helped to vindicate. Meanwhile, there was plenty within

4. Cf. his letters to Richard Bentley, and the General Scholium added to the second edition of the *Mathematical Principles*, pp. 543–7.

physics to keep him busy—plenty of genuinely scientific questions, which one could hope to answer by reference to the telescope or an experiment. He had no time or inclination to defend the notion of gravitation from other people's misinterpretations.

From this example we can perhaps see what is liable to happen when scientific terms are used, not to explain anything, but for other purposes—for instance, as the raw material of myths. Technical scientific notions taken by themselves have, as we saw, about as much meaning as isolated pieces taken out of a jig-saw puzzle. If we try to do other things with them—for example, to build a comprehensive 'world-view' of a philosophical kind from them—we are forgetting this fact, and treating them as though they were pieces of a single, cosmic jig-saw. This has two unfortunate consequences. First, you cannot get pieces taken from different puzzles to fit together at all except by distorting them; and in the second place, if one man forces them together in one way and one in another, nobody will be able to say that one or the other of the pictures so produced is, scientifically speaking, the 'right' one.

These difficulties arise again when physical or biological theories are appealed to in an attempt to solve problems in, for instance, ethics or political theory. To begin with, all the scientific terms used get distorted in the process, and no longer keep the clear meaning they have in science proper: this fact alone shows the gulf between scientific myths and the theories whose concepts they exploit. Furthermore, when two people appeal to the same scientific theory as backing for different 'world-views' or different political doctrines, how can we even set about choosing between them? Within science, we can at any rate prove our views in practice. But when we put scientific terms to non-scientific uses, this, the chief merit of a scientific approach, is lost. For all that experiment or observation can show, one scientific myth is as good as another.

The Limits of Cosmology

Men have always been curious to know about the Beginning of All Things and the End of All Things, so mythologies have at all times contained Creation stories, and often Apocalypses too. We are beginning to realize now what complex things these myths are: how many strands of thought are entangled together in them, and how mistaken it is to suppose that they represent only attempts to say what happened a very, very long time ago and more, and what is going eventually, eventually to happen. But that they have represented these among other things is past question, and it is worth unravelling this strand from the others and looking at it by itself. So let us take the questions 'What happened before anything else happened?' and 'What is going to happen after everything else has happened?' and see what we can make of them.

We certainly cannot expect these questions to be easy to answer. If scientists have any contribution to make to their answering, we must presumably expect it to be a tentative one, for how could it be anything else? If the questions are meant as historical ones, they must be tackled as such; and there are certain difficulties about any sort of historical study, whether it is the diplomatic history of Europe during the last quarter-century, the archaeological reconstruction of a civilization dead for three millennia, the geological story of the formation of the coal-measures, or the astronomical history of the solar system. In each case we must expect the more remote past to be shrouded in greater mystery than the less, and we cannot hope, short of great good fortune, to reach the same certainty about the remoter past that we can about the more recent.

Strokes of luck do occur sometimes, such as the preservation in peat of the lake-village at Glastonbury, thanks to which we know a good deal about life there during the last centuries B.C., though the early centuries A.D. are still in darkness. But by and large one can fairly say that the further back we go the less we can hope to find out. What goes for the reconstruction of the past holds with greater force, if that be possible, for the prediction of the future. The longer the term of the predictions we make, the more qualified they must be, and the same darkness that envelops the remoter past falls even more quickly when we look into the more distant future.

This at any rate must be true of any attempt to argue methodically from evidence about the state of things to-day to conclusions about the state of things a long, long time ago or far, far into the future. We do not find, and perhaps we do not expect to find, the same modesty about Creation Myths. It may be that, intellectually, the Book of Genesis is the worse for lacking these qualifications: this I am not going to discuss. What is clear, however, is that a scientific account of these things must certainly not lack them: '. . . And the Earth, so far as we can ascertain, was void; and the evidence tends on the whole to support the view that darkness was upon the face of the deep . . .': this may seem pedestrian, but it is the business of scientists to be pedestrian, to keep one foot on the ground of their evidence, and not to run, leap or vault off into unsupportable speculations.

With these cautions in mind, I want to turn and look at the things scientists have recently had to say about the remotest past, and about the remotest future. The cautions are necessary, for in each case one finds their utterances strangely confident and un-tentative. If things had worked out as we should have expected, nothing in science would have been less certain than our speculations about the very beginning of all things, and about their ultimate fate. Yet when scientists turn to discuss these topics, a sudden fluency descends upon them, as though the mist through which the past and future are seen was, when we reach the extreme limits, suddenly lifted, and a clear vision of the first and last events granted to us. What has happened? Is it that, like the archaeologists of Glastonbury, students of physical cosmology have been blessed with unexpected good fortune, in the shape of striking and conclusive evidence in a field where they could hope for no more than ambiguous scraps? Or has their confidence a different source—is their determination to get at answers to these questions affecting their sense of relevance, and leading them to find evidence where there is none? Here is a question at any rate worth posing.

Let us start with the stories about the uttermost future: in particular, with a doctrine which is a common-place of popular scientific addresses, the doctrine of the 'running-down universe'. This is something which hardly needs expounding in detail: most of us must have come across one or another of the apocalyptic utterances in which physicists, philosophers and theologians discuss the ultimate 'death by freezing' of the entire universe. The central suggestion, it will be remembered, is this: that one of the best-established laws of physics, the one called the Second Law of Thermodynamics, obliges us to think of the universe as a one-way system bound of necessity to become uninhabitable, that—saving some sort of divine intervention—all activity in the universe is as certain to peter out as a clock which no one winds is bound to stop. Though the exact date at which life as we know it will be extinguished may be a matter of doubt; though the time involved may be measured in millions of millennia; nevertheless, when the temperature drops far enough, the end must come, and life will be over for good. The running-down of the universe may be slow, but it is (we are told) inexorable. Whatever else may be uncertain about the future, that at any rate is in store . . . and after that, nothing. This much we know for sure; and all that there is left to us to do is to face the consequences of the discovery.

This conclusion has always seemed a bit of a nightmare, and just how we should compose ourselves in the face of it people have not been able to agree. Philosophers like Russell and Ramsey, theologians—Dean Inge being an example—scientists such as Ostwald, a Nobel Prize chemist: writers of all kinds have concerned themselves with its implications.[1] Recently, too, Fred Hoyle has even argued that, as a matter of physics, the Second Law may not be universally applicable after all, that processes may be going on in

---

1. F. P. Ramsey, *The Foundations of Mathematics*, p. 291; W. R. Inge, *God and the Astronomers*; F. W. Ostwald, *Die Philosophie der Werte*, p. 98.

the universe which are capable of making up for the Universal Unwinding and putting fresh power into the Spring, so that we can breathe freely again. But for once let us leave the details of what they all say, both the physics and the attitudinizing, and see whether we cannot get behind the dispute, asking: Was there ever any real cause for a nightmare? Whatever the scientific rights and wrongs of Hoyle's new theory, need we ever have felt worried in the first place? Could any discovery physicists might make ever in fact compel us to regard the universe-as-a-whole as a 'running-down clock'?

There is no denying that, if we are in fact forced to accept this picture of the universe, if physicists can really predict the eventual extinction of all effective temperature-differences, and if this prediction is as utterly and absolutely inescapable as we have been told, then—even though the moment of doom may be millions of millennia away—this must make a considerable difference to our view of the world. There are, no doubt, several possible reactions. One can be heroic about it, like ocean voyagers who continue to dress for dinner even though they have discovered that the ship is sinking under them, feeling that there is a certain dignity in putting on a good show in the meantime. One can be Epicurean, like Frank Ramsey, who called for a sense of perspective: 'In time the world will cool', he wrote, 'and everything will die, but that is a long time off still, and its present value at compound discount is almost nothing'. Or again, one can be other-worldly: if the scientists now confirm the view that the world of space and time is a leaky vessel, that (one can argue) shows all the more that we should pin our hopes Elsewhere. But whichever our reaction may be, whether we choose Jeremiah, Epicurus or Casabianca as our model, we must do something to reconcile ourselves to the inevitable future. As Ostwald put it: 'We must in all circumstances learn to accept the fact that at some indefinite but far-off time our civilization is doomed to go under . . . and that, in the longest run, the sum of all human endeavour has no recognizable significance.'

But all this is conditional. And the conditional clauses are these: *if* the picture of the universe as a running-down clock really has all the authority and backing that it seems to; *if* it rests on a proper reading of the Second Law of Thermodynamics; *if*, when we are told that it is theoretically impossible to decrease the entropy of an isolated system, this impossibility is of a kind that calls for regret or resignation or fatalism. These will turn out to be three very sizeable 'ifs'.

At this point, now the word 'entropy' has appeared and needs explaining, an excursion in the direction of physics is unavoidable. I have also quoted the standard formulation of the Second Law of Thermodynamics—namely, that in a thermally isolated system all physical changes take place in such a way that the entropy remains constant or increases—and it is necessary to see what this law implies in a practical case. These two explanations are best given together. So, by comparison with our earlier example, we may perhaps say this: as the term 'gravity' is introduced for purposes of mechanics, so the term 'entropy' is introduced for purposes of thermodynamics, that is, the theory of heat-exchanges. Where the law of gravitation is used in the first place in explaining, in a mathematical way, how the planets, the comets, falling bodies and the waters of the oceans may be expected to move, the Second Law of Thermodynamics is used in the first place in working out what efficiency you can expect to get from a steam-engine operating at a given temperature, how much power will be needed to run a refrigerator under given conditions, and so on.

'In the first place', we have to say, as the law soon acquires extensive affiliations of a more abstract and theoretical kind. These we can look at later: it will be as well to start by considering the law in its simpler, 'phenomenological' form, in which its cash-value in terms of actual happenings is more easily grasped. In practical terms, then, the force of the law is, that if a system of bodies is shielded from exchanges of heat with surrounding bodies—completely lagged, that is—the temperatures of the various bodies in the system will tend to even out, the hotter ones cooling and the colder ones warming up. Just as we have the technical word 'temperature' as the numerical counterpart of the familiar words 'hot' and 'cold', so we have the technical word 'entropy' as the numerical measure of the degree to which this evening-out process has gone on. The exact manner in which entropy is measured we need not enter into. What we do need to notice is that, at the phenomenological level, the law has to be stated in terms of 'thermally isolated' systems. If we are to know how far the law is applicable to any chosen system in nature, we must first inquire how far it is shielded from thermal interaction with its surroundings; roughly speaking, how far it is lagged.

How are we to draw from a law of this kind any conclusions about the universe as a whole? Can we do so, indeed? From the start we saw grounds for suspicion. How physicists feel so sure about what is going to happen all those millions of years away must be a bit of a mystery. If, like the astronomers in H. G. Wells's story *In the Days of the Comet*, they had the best of evidence that another vast heavenly body was about to collide with the earth, things would be rather different. That discovery would be something like a 'death-sentence': we should have some reason then to start saying our prayers. So the first thing which was fishy about the 'running-down universe' argument was the indirectness of the experimental evidence. If an astronomer were to warn us of the imminent impact of a comet, we could at least ask him for direct evidence, such as the observed trajectory of the comet over the past weeks and days. If the ultimate fate of the universe were predicted on the basis of this sort of argument, we must therefore ask: what sort of evidence is relied on? In the first place, observations on the performance of steam-engines. . . . To which the layman might be tempted to reply, 'Quite: and the Roman augurs used to predict the fall of cities from a study of the intestines of birds.'

In fact, this comparison is far from just; and besides the crucial objections to the argument lie elsewhere. It is by a more complex train of reasoning that the ultimate freeze-up is predicted, so that it is not enough to complain about the indirectness of the evidence. In any case that is not the target on which we should concentrate our fire, for, so long as we feel only that the evidence is at fault—so long as the clock picture itself is left uncriticized—it will be natural to suppose that, given time to collect more evidence, physicists may yet succeed in establishing the doctrine. At any rate, the idea will not seem absurd.

This, however, is what really needs examining: whether physics in general or thermodynamics in particular can have this sort of implication at all. To ask this question is not to cast doubt on the acceptability of the Second Law as a law of themodynamics. (This, to repeat, I am regarding as unquestioned.) It is to ask rather, as a logical matter, whether to appeal to the law in support of metaphysical doctrines in the sphere of eschatology is not to misapply it.

Perhaps putting the issue in this way begs the question, by implying that the doctrines we are concerned with are not physical but metaphysical ones. This certainly needs to be established, since a great deal turns on just this fact. So I want to argue that the thesis which likens the universe to a running-down clock is a double one, one half genuine physics, the other half metaphysical, and to show what is involved in assuming that both halves can be established by the same scientific methods.

We are told (it must be noted) not only that our earth will eventually cool down to an unendurable extent, and all life be extinguished, but also that this is only one aspect of the inexorable decay of the whole universe. If the first half of the thesis alone were advanced, the issue would indeed be largely one of physics. We can certainly visualize life on the earth coming to an end if all the regions at present inhabitable were to freeze up, and, if this were all that was claimed, we should be faced with a straightforward prediction. Still, it is a prediction which could be made with confidence only on the most unlikely assumptions. After all that men have managed to achieve during the last few hundred years, are they going to sit by and let themselves be gradually snuffed out, over a period of millions and millions of years? It needs little enough imagination to suppose their finding ways of keeping up the surface temperature of the earth, developing in their descendants greater resistance to cold, if need be jet-propelling the orbit of the planet a little nearer the sun. There are countless things they might do to falsify the prediction, many of them at the moment no more than Science Fiction, but none of them out of the question if you consider the amount of time at their disposal. As a practical threat to the future of the human race, the cooling of the earth cannot rate very high; we can all of us name several far more serious.

This reply, however, deals only with the first half of the thesis, and if we stopped at that point we might be accused of missing the real significance of the Second Law. 'Any things which men did in this way', it might be said, 'would be merely palliatives, and could only postpone the end. It is not as an immediate, practical obstacle that the effects of the Second Law are important: their true significance becomes apparent only when you widen your vision to embrace the whole universe, and when you realize that to counter the action of the law is more than a practical problem—it is a theoretical impossibility.' 'The second law of thermodynamics states' (and now I am quoting Sir Harold Spencer-Jones, the last Astronomer-Royal) 'that entropy in the universe must always increase; so that all change will eventually cease, and this ending will come in a finite time—the universe is running down and must eventually come to a stop—and this law, so far as we can tell, holds a supreme position amongst the laws of nature.'[2]

Here the second half of the thesis begins to come into play, and the nightmare impression of fatalism begins. So long as we thought only of the earth the problem seemed of manageable size, and so long as it was a practical problem we were tackling there seemed hope of finding practical means of getting round it. But the problem cannot be got around, it is now argued, as what we are up against is a theoretical, not a practical barrier; and further it is not the solar system alone but the whole universe which is grinding to a stop, so that all our palliatives will in the long run be in vain. However, it is at this point also (so far as I can see) that the issue ceases to be a genuinely physical one. Suppose it is completely established that the Second Law can be applied to all physical systems thermally isolated from the rest of the universe, does it necessarily apply also to the universe taken as a whole? And does the fact that the impossibility it tells us of is a theoretical rather than a practical one imply that it is exceedingly, indeed infinitely, difficult to overcome?

These are the propositions we must question, for two reasons: both because their soundness is regularly taken for granted when this subject is under discussion, and because it is only by introducing them into the argument that the clock picture can be established—without them, the fatalist *Weltanschauung* shrinks into a far-distant challenge to the technology of our descendants. We can concede that the Second Law has won for itself a supreme position in the structure of physical theory, and that it is now accepted as a

2. *Science News*, No. 32 (Penguin Books 1954), p. 24.

'universal law' which expresses a 'theoretical impossibility': what we must examine is the further inference that the law tells us something ineluctable about the universe-as-a-whole. So let us look at the two phrases 'a universal law' and 'theoretically impossible', and see what traps they hold in store for us if we are not on our guard.

First, then, the phrase 'a universal law'. To say that the Second Law of Thermodynamics is a universal law is to say that it holds *universally*, in the same way as the law of gravitation. (Of course, since Fred Hoyle put forward this theory we are not so sure that it does in fact hold universally after all, but this is by the way: for our purposes it is better to suppose that the universality of the law is established and ask what follows from that.) A universal law, in other words, is one which has been found applicable not just to some but to all systems of physical objects which satisfy the conditions laid down in the theory. The law of gravitation is universal if and only if it holds for all pairs of objects having mass; and the Second Law of Thermodynamics is universal if and only if it holds for any thermally isolated system, whatever and wherever it may be—and holds the more nearly for any system we choose, the more completely this system is thermally isolated from its surroundings. By itself, the fact that a law is a universal one implies nothing about the universe-as-a-whole. The fact that the law of gravitational attraction held universally would never be taken as implying that 'the universe' must be attracting something, any more than the discovery that tooth-cleaning was a universal practice would imply that 'the universe' must clean its teeth. A statement which 'holds universally' is one thing, a statement about 'the universe' is another, and a step from one to the other will always require justification.

In the case of gravitation this step is never taken: in the case of entropy it is taken regularly, but the justification is assumed—indeed, from the way in which it is taken one would not guess that any question of justification arose. According to the late Astronomer-Royal, for instance, the Second Law 'states that entropy in the universe must always increase', and that is that. But this is a sheer mis-statement of the law. The Second Law by itself states nothing about 'the universe', any more than does the law of gravitation.

The most it could do would be to *imply* something about the universe, and it could do that only if we also knew how far the universe was itself a thermally isolated system. Until this additional question has been squarely faced, we cannot be satisfied that there is any warrant in physics for the idea of the whole universe coming to a stop.

How could this further question be settled? Is it, in fact, the sort of question a physicist could ever hope to answer? One of two things seems to be needed: either scientists must make direct observations to establish the universe's degree of isolation, or they must find some way of proving from other established facts what the extent of its isolation is. But once we start to imagine possible experiments, and inquire what observations would really serve us here, we come up against grave difficulties. These are not the practical difficulties, of making observations on so vast an object of study. They are conceptual or (if you prefer) intellectual difficulties, to do with the sense of the question itself. For the question is, how far the universe-as-a-whole is a thermally isolated system, a 'thermally isolated' system being one which is shielded against all interchanges of heat with bodies outside itself; and what are we to make of the question, whether or no the universe-as-a-whole is shielded thermally from its surroundings? Since every material system and part of space there is forms (by definition) a part of the universe, we have left ourselves no room to talk about the universe's having surroundings. It is not just that outside the universe there is nothing, so that from that quarter it is incapable of drawing any fresh supply of heat: it makes no more sense to talk of the surroundings of the universe as empty than it does to talk of them as full, for in this context the distinction between 'inside' and 'outside' has no use. The prime difficulty lies deeper: it is that, whereas the question how far a given physical system is isolated from its surroundings has a clear enough meaning when asked about any bounded part of the universe—being equivalent to the question, to what extent heat exchanges are possible across the boundary—when asked about the universe-as-a-whole, its meaning is completely obscure.

It has long been notorious that questions which can be asked with perfect propriety of particular things, or parts of the world, or stretches of time, tend to go wrong on us if we ask them about 'everything-there-is', or about 'the universe-as-a-whole', or about 'time itself'. This happens, for instance, with questions about 'the beginning of time', and it happens again in our present case. Talking about 'shielding the universe from its surroundings' has no more literal significance than talking about 'frying minus three eggs'. So unless some other more intelligible sense can be given to the question, the conditions necessary for us to apply the Second Law of Thermodynamics to the universe-as-a-whole are such as *cannot* be satisfied.

No alternative interpretation is normally offered, presumably because the need for it is not seen; and, though the hunt for suppressed premises is always a speculative one, it is tempting to conclude that among them is an uncritical identification of 'universal laws' with 'laws of the universe'. Without this, the story that the universe is running down loses a good deal of its bite.

All this argument has, it is true, been presented in a form which applies to the Second Law in its simplest form—the one in which it applies most directly to the happenings in the world. Will the situation be altered if we treat the law differently, and consider it in some more abstract and theoretical form? 'Surely', it is said,[3] 'there are half a dozen alternative formulations of the law which do not mention "thermal isolation", and which can therefore be applied to the universe without assuming that it is "lagged" against something outside. If these formulations are equivalent to the original one, that shows that the clock-picture is valid after all.'

It would be a long and technical job to consider these alternative ways of stating the law one by one. In any case, two considerations may help us to see what the result would be. First, if the other formulations are truly equivalent to the original one or, being more general than it, reduce to it when one is considering thermal phenomena alone, then it is difficult to see how they can be in any different case when we come to apply them to 'the whole universe': if two questions are equivalent, a logical incongruity in one will be present in the other also. But secondly, the presence of the incon-

3. E.g. by Professor M. Polanyi in a comment on an earlier version of this essay (*Listener*, 15 Mar. 1951, p. 423).

gruity may be concealed, if the terms used are of an abstract and theoretical sort, for we may then fail to examine carefully enough the steps which must be taken if the law is to be used to account for actual physical phenomena.

This will be particularly liable to happen if the terms used are not obviously recondite, like 'entropy', but are borrowed from familiar speech and so apparently innocent. It may, for instance, be suggested that what the Second Law tells us is that 'order' is always lost in physical processes; and that to say the universe is running down is to say that it is, so to speak, like a vast box full of packs of cards which, when shaken, lose their original order and become more and more mixed up.[4] But it is a long road from the theory of statistical mechanics, in which such analogies as this are at home, to the phenomena the theory is used to explain. How far the analogy between 'the whole universe' and a box full of cards can be pressed is something we cannot discover by theorizing alone, but only by considering in detailed, practical terms the relation between the theory and the facts it explains. For this purpose, the more abstract the formulation of a law we consider, the more confusing it will tend to be—especially if the terms in which it is expressed are deceptively concrete and familiar.

Appeals to abstract theory are, in any case, liable to cut both ways, as we can see if we turn to the second of the two notions that we set ourselves to examine. This was the notion of 'theoretical impossibility'. Seeing that this notion is what gives the clock-picture its air of fatalism, and is the source of the idea that the universal freeze-up is not just a threat, but an utterly inevitable doom, we had better take a closer look at it too.

When we read about 'theoretical impossibilities' in physics, we naturally think of them as things which are tremendously hard to do, and more. Lifting a ton weight single-handed—that is something no man yet born has been strong enough to do. But decreasing entropy, or cooling things below the Absolute Zero, or getting around Heisenberg's principle: these we understand to be harder still, still more obstinate—things in fact which there is utterly no hope of doing. To overcome a theoretical impossibility, we feel,

4. Cf. R. E. D. Clark, *Listener*, 15 Mar. 1951, p. 423.

we should have to surpass even the U.S. Transportation Corps, whose motto claims that they do the difficult every day, while 'the impossible takes a little longer'. Theoretical impossibilities seem not just harder than practical ones, but infinitely harder. This is one reason why the Second Law of Thermodynamics has made people feel fatalistic.

Yet is the difference between practical and theoretical impossibilities of this kind at all? There are strong grounds for thinking not, as may be seen if we look at a less technical example. Some things are, practically speaking, impossible to weigh; but, with others, weighing is theoretically impossible. We may find it hard to weigh a snowflake on a pair of kitchen scales, so hard as to be, practically speaking, impossible; but if someone says to us 'You can't weigh *fire*' the impossibility is more than a practical one—it is a theoretical one, which there is no question of our getting round by improving our instruments or increasing our ingenuity. This is not to say that weighing fire is that much harder again even than weighing a snowflake: it is to say something quite different. For what makes it impossible is not our want of ingenuity, or of suitable apparatus: no increase in skill or progress in instrument design could affect the issue. No: the root of the impossibility lies in the fact that our system of chemical classification or concepts does not admit 'fire' as a kind of substance.

Once this was not so. Only since the end of the eighteenth century has the distinction between chemical substances, physical processes and states of matter been clearly established. Before that, either 'fire' itself or phlogiston, the 'fiery element', was placed alongside water and air in the categories of science. Nowadays, however, the categories of chemistry are more refined. The process of combustion, the products of combustion, energy of the reaction, flaming gases and objects burned are all logically distinguishable; and the term *fire* has in consequence become the name of a process, not a substance. As a result, though questions about the weight of fire once had some sense, even if an obscure one, now they have none. Now we can no longer 'weigh fire', any more than we can 'weigh compassion' or 'weigh Tuesday'. Anything which it makes sense to talk of our weighing we may hope some day to weigh—including such things as the burning gases in a flame. Only if we chose to use the phrase 'weighing fire' to mean, say, weighing flames could the question of its possibility be revived.

This example may seem to turn on a matter of words—though that impression would be a great mistake, since it is chemical discoveries as enshrined in chemists' categories that concern us here, rather than just the words they use for marking their distinctions—so a rather different example may help to bring out the point. A man who sets out on a detailed and exhaustive survey of the Polar regions will encounter all sorts of practical difficulties. Some of these are so great as to make the task at present an impossible one to complete; still, there is room to talk of finding a way of doing it some day, given suitable equipment and sufficient grit and determination. This is a nice example of a practical impossibility, the sort which might one day be, but has not yet been, overcome. Someone, however, may point out that the task involves further difficulties and impossibilities which no such equipment and determination can get over. 'Suppose you have been able to make all the observations you require,' he may say, 'even then your difficulties are not over. For, adopting Mercator's projection, you will still not be able to produce your map. Just as Baffin Land comes out large and Greenland larger still, so each mile further north you survey will take up more paper; and, however large you make your map, you will never be able to show the North Pole on it. And this is no mere practical obstacle which we may hope to get over some day: it is a theoretical impossibility, and there's no room to say, ''Perhaps some day we'll do it''.'

Still, this impossibility is again nothing to worry about—nothing 'unpleasant' that needs 'facing'. For if we chose we could map the same region to a different projection, and then the North Pole would appear on the map in the same way as any other place on the earth's surface. In this case, therefore, the theoretical impossibilities are not more difficult, not even infinitely more difficult to overcome than the practical ones. To speak of difficulty here at all is to misconstrue theoretical impossibilities as a specially obstinate variety of practical ones; whereas all talk of fatalism, of hope deferred, vain attempts or resignation would be a sign of misunderstanding. A theoretical impossibility is not a challenge, so philosophical attitudes are out of place.

What application has this argument to the story of the Running-Down Universe? The suggestion was that the Second Law of Thermodynamics had apocalyptic implications for two reasons: first, because it applied to the whole universe, and secondly, because it represented a theoretical, and so ineluctable necessity in events. But even if, given its more abstract, theoretical interpretation, the Second Law were applicable to the universe-as-a-whole, would it yet have the consequences depicted? Only if the necessities and impossibilities it told us of were (as is implied) infinitely hard practical ones. This assumption, we can now see, is also highly questionable. Principles of this degree of abstraction may be expected to belong to the conceptual scaffolding of a scientific theory, and the necessities and impossibilities they state will be (so to speak) built into the theory. The phenomena being what they are, we have, no doubt, built up the theories we have for very good reasons. But to say, for instance, 'Processes cannot be weighed, substances can', is not to state these reasons: it is to presuppose them.

It seems fair to suggest that the same conclusion applies to the Second Law itself. There is one crucial test: if there is anything in the implied parallel between theoretical principles of physics and cartographical principles of projection, we shall be free to shift from one set of principles to another at will, provided only we are prepared to accept all the consequential changes in the details of our theories and concepts. So it should be possible, in principle, to adopt in place of the Second Law another law which does not seem to imply a permanent long-term trend in the course of the universe's history: this would be like shifting from Mercator's projection to another projection—one that does not seem to imply that the North and South Poles have a peculiar status, which sets them apart from all other points on the earth's surface, viz. that of unmappability.

Can this be done? One thing has to be said at once: that working physicists have no reason to be dissatisfied with the existing structure of thermodynamics, and so have had little motive for investigating this question. There are no phenomena which a revised thermodynamics would accommodate which cannot be explained within the existing framework of theory. (The reason for choosing one map projection rather than another is not usually that one projection leaves unmapped areas that can be mapped on the other.) The question has, however, been studied by Professor Herbert Dingle, who concludes that the transformation is possible; the laws of heat transfer can, with sufficient ingenuity, be framed in an alternative way which does not involve any suggestion of 'universal thermal decay'. What is perhaps significant is that, if this is done, we can unify heat theory and dynamical theory only by introducing into mechanics itself something akin to the existing entropy law.[5] One can, it seems, start constructing physical theory either with mechanics or with heat theory: whichever way one starts, one will be free of any apparent implication of decay, until one crosses into the other subject. How can it, then, be argued that universal decay is a theoretical necessity inherent either in thermal or in mechanical *phenomena*? Surely the more natural conclusion to draw is that the source of the so-called 'universal tendency to disorganization' lies elsewhere—perhaps in the analogies we have to employ in order to accommodate either the concepts of thermodynamics in a generalized theory of mechanics, or those of mechanics in a generalized theory of thermodynamics.

Let me sum up this long argument. People have thought that we could read the Second Law of Thermodynamics as telling us about the ultimate fate of the universe. So long as physicists accepted the law as a theoretical principle of universal application, it seemed to follow that the universe must be running down. This being so, it was felt (understandably) that we must do something to reconcile ourselves to the ultimate doom; though whether we were to take it heroically, cheerfully or in an other-worldly spirit people could not agree. But the argument is open to criticism on two counts. If we take the Second Law at the phenomenological level, at which it relates most directly to the facts of heat-transfer, the argument works only if we assume that a 'universal' law necessarily applies to 'the universe'—and there turn out to be strong reasons for thinking that the law could not be 'about the universe' in the sense required. Alternatively, if we give the law its most theoretical interpretation, the argument works only if we assume that theoreti-

5. Herbert Dingle, *Through Science to Philosophy*, Ch. 11; and also his contribution to the symposium, *Albert Einstein, Philosopher-Scientist* (Library of Living Philosophers), Ch. 20, and *The Scientific Adventure*, Ch. 16.

cal impossibility is a sub-species of practical impossibility (namely, the *infinitely* difficult): it is this which gives the Second Law its misleading air of inexorability. But so far as an impossibility might (conceivably) be overcome, it is a practical one; and so far as it is a theoretical one, it has no need of being.

All this is not to say that philosophical attitudes to the universe-as-a-whole are unjustifiable, and cannot be argued about. Perhaps we *should* be stoical about the ultimate fate of all things, or carefree, or other-worldly; and no doubt reasons of some sort can be given in favour of adopting one of these attitudes or another. All we are entitled to say is, that physics does not *oblige* us to adopt one, or indeed any, of these attitudes. The running-down universe is a myth, and we shall discover about the Apocalypse from physics only what we read into the subject.

The pitfalls surrounding the notion of an Apocalypse surround equally the idea of a Beginning of All Things; and if we force our astronomical discoveries and physical theories to tell us about this, we once again run serious risks. It is not that we are liable, in so doing, to take for the truth of the matter propositions which are in fact untrue. The prior and greater risk is that of taking for useful and sensible hypotheses suggestions which turn out on closer study to be vacuous or incoherent.

The situation is complicated for us by the fact that we have some evidence of a genuine long-term trend in the progress of past astronomical history. Though the observations are open to interpretation in a number of ways, it does seem quite likely that the galaxies are getting farther apart from one another, and that if we went back far enough in time they would have been very much closer together than they are now.[6] This conclusion has led some people to make suggestions which are not so much extravagant as unintelligible. We are asked, for instance, to suppose that, when the galaxies which are now separate were all packed into a relatively small space, there was no change, and so, 'if time has a meaning only when there is change', no *time*. 'We can conceive', says the late Astronomer-Royal, 'on one view, of a timeless past while creation slept and nothing changed, and on another of an

6. Cf. G. J. Whitrow, *The Structure of the Universe* (Hutchinson 1949), Ch. 2.

initial creation and of a beginning of time only a few thousand million years ago'.[7] Now there may undoubtedly be difficulties about applying our familiar conception of time at all straightforwardly to a universe as different from ours as that the astronomers think may have existed a few thousand million years ago—though these difficulties are apparently not serious enough to stop them putting an approximate date to it. But the notorious objections even to *asking* when time began are unaffected by anything the telescopes may show us, and the 'two views' we are offered remain equally mysterious.

Can the views even be kept distinct? It is not clear that they can. They are alleged to differ in this, that the one supposes the phase at which the galaxies were most highly compressed to have followed 'a timeless past when nothing changed', while the other supposes it to have followed . . . nothing. But to talk of a genuinely 'timeless' past, a past 'when time had no meaning', would be to talk in a way which deprived of sense even the proposition that *nothing* changed: if the assertion 'Nothing changed' is to be intelligible, must one not at any rate be able to ask *for how long* this state of affairs continued? So we may wonder whether the choice between 'a timeless past' and no past at all is a genuine choice after all.

Either way, whichever 'view'—or better, whichever form of words—we favour, violence is done to our understandings. We might make something of the hypothesis that, up till a few thousand million years ago, all the matter composing the present galaxies had remained stationary and compressed together for an indefinite time, though what evidence would be required in order to establish that this had been the case is not obvious—certainly none that astronomers now possess would be enough to do so. But to present this indefinitely long unchanging phase as either 'a timeless past' or as 'before time began' is to play, not for understanding, but for mystified acquiescence. Perhaps the phase of maximum compression was the beginning, not only of the current phase of astronomical history, but also of . . . all objects and happenings whatever. But this is hardly the sort of thing which could be proved by resort to the telescope: an event which was logically as unique as a

7. *Science News*, No. 32, pp. 23–4.

Creation of All Things would not be open to discussion in terms of ordinary induction and analogy. As things stand, we have no reason to suppose that the phase of maximum compression did not have predecessors in a perfectly familiar sense—for all we now know, there may prove to be heavenly bodies, sufficiently distant, which were not involved in the initial 'jam' of galactic matter. If we are to believe anything else, indeed, it can hardly be for astronomical reasons alone. One could not seriously identify this 'jam' with a Creation unless one had outside reasons for thinking that this was justifiable.

What sort of outside reasons could there be? Professor Milne, whose theories Dr Spencer-Jones was popularizing in the passage just quoted, was greatly interested in the question what mathematical 'time-scale' one should use in theorizing about the past history of the galaxies. One of his suggestions was that one might use two alternative scales: one of these would be like the Orthographic map projection, or the physicist's 'absolute' scale of temperature, in being bounded in the direction of the past—on this scale, that is to say, the units would be such that only a finite number of units back from now would be properly describable as past 'times'; the other scale would be more like Mercator's projection, or the 'logarithmic' scale of temperature, in being unbounded—on this alternative scale, any number of units back from the present time would represent a 'time' in the past. So far as pure theory goes, there is no reason why such alternative scales should not be used: difficulties begin only when we try to draw conclusions from this fact. For the choice of a time-scale, in this sense, is a conceptual matter alone. We may make our choice in the light of our discoveries about astronomical history, but the choice itself tells us nothing about history: it is a preliminary matter, like the choice of a projection in cartography. What scale we choose makes no difference to the events we have to chronicle; and the suggestion Milne makes, that the moment of greatest galactic compression might be one and the same as the initial point of his bounded time-scale, is no improvement on the bare hypothesis that this was 'when time began'.

At the present stage in physical cosmology, it is no good our being in a hurry. As in history and archaeology, the only hope of solid results lies in a step-by-step advance. There will always be some point in our map of the past history of the universe beyond which we have nothing particular to show. The same goes for the future—and how could it be otherwise? If we force astrophysics to serve us with a revised version of Genesis and Revelation we dig a pit for ourselves. Suppose we look at the map of an unsurveyed country, we shall find that it is not completely empty, for it will bear at least the parallels of longitude and latitude. But it would be a mistake to think that these, by themselves, told us anything geographical about the surface of the earth; and the same pitfall awaits us in cosmology. However we may argue about our choice of time-scales, these will remain the bare scaffolding of astrophysical theory. The danger is that we may misinterpret them as giving us genuinely historical information.

Ethics and Cosmic Evolution

The myths of the past have always had justificatory purposes as well as explanatory ones. They have set out to show people not only how it is that Nature and Society and Man have come to be related as they are, but also why these relations are rightly as they are. To the peoples of the Ancient Near East, a royal genealogy was more than a historical record, it was a warrant—a guarantee of the legitimate authority of their rulers—which related the order of society to the order of nature, usually by crediting the sovereign with a direct line of descent from the creator-gods of the country. The mythical figures of their cosmologies, again, stood in social relationships to one another as much as men did; and the pattern of these relationships commonly reflected the social structure of the community itself, so that it seemed in the nature of things that the social structure should be as it was. Ethical and political issues were thereby given cosmological foundations, the Nature of Things being identified with the Right Order of Things.[1]

This ambition, to find 'a cosmic sanction for ethics', a 'natural foundation on which our human superstructure of right and wrong may safely rest', is an enduring one. We feel there should be some sign in the world of nature which can reassure us that our ideals of society and morality, as they have grown up through the centuries, are enduring and of real worth. Can we not find some parallel between the pattern and progress of society and their counterparts

---

1. H. Frankfort *et al.*, *Before Philosophy* (Pelican edn.), give a useful account of this subject.

in the rest of nature? Can we not discern some order and direction in the development of things at large? This is a problem any man may think worth tackling.

Scientists have naturally been concerned with this problem as much as anyone else, and have often thought that the results of their professional inquiries could be relied on to provide a solution. To take one example: Aristotle introduced, for purely biological purposes, the idea of a 'scale of nature'—in this scale all things could be ordered, rocks at the foot and men at the head, and each upward step took one farther away from the inert and towards fully living and reasoning creatures. His one point in constructing this scale was to emphasize the continuity between living things and inert ones. Wherever one thought to draw a sharp line across the scale, borderline creatures could be found, so that stones, plants, animals and men could not be classified in entirely separate pigeon-holes, rigidly insulated from one another: there would always be such things as sponges (or in our day viruses) which would blur the divisions in any attempt at a rigid classification.[2] In later times, however, this idea of Aristotle's was given quite a different interpretation. From being an account of the way things were *in fact* related, it became a specification of the hierarchy in which they *ought* to be related. The Scale of Nature became the 'Sovereign Order of Nature'; and now a creature's place in the scale—the extent to which it differed from the entirely inert and approximated towards the ideal of full life and rationality—was made to represent also its 'station' in the Society of Nature. A creature at one level was given 'authority' over those at all lower levels, and was subject in turn to all those above it. Man being at the head of Aristotle's scale, he was set in dominion over the lower creatures; but now Angels were placed above him in the scale, with God in the topmost place of all, man being as much a subject to them as the lower creatures were to him.[3]

Whatever sympathy one has with this view of things, the view is more than a biological one. If we are going to order the creatures of nature in a hierarchy, as opposed to a strictly taxonomic scheme,

2. C. Singer, *A History of Biology* (Oxford 1931), Ch. I. §11.
3. E.g. S. Mason, *A History of the Sciences* (1953), pp. 141–2.

this may be as good a way of doing so as any; but the question whether it should be done in this way or some other is not a question which can be determined from biological considerations alone. Suppose someone were to argue, as against this view, that authority in the world of nature should derive from longevity rather than degree of life or rationality: as men among themselves regard their parents and grandparents as entitled to honour and respect, he might say, in the same way all men, as comparatively short-lived creatures, should honour and respect the tortoise, the carp and the elephant. A hierarchy constructed on this principle would differ strikingly from that based on Aristotle's scale—for instance, it would make the rocks of the Western Highlands the most honourable things in the British Isles—yet how could one choose between them? Not on biological grounds alone surely: neither view could be claimed as 'biologically correct' or dismissed as 'biologically unsound'. So if the disputants turned to a biologist to arbitrate between them, he would have to reply as follows: 'In this matter I cannot claim to be an expert. As a biologist I can tell you what sorts of creatures there are, and in what ways they resemble and differ from one another; I can order them for you on different principles, and tell you what principles of classification are the best for the special purposes of my science; but when it comes to what you call "authority" and "respect", you leave my province. The question whether longevity is the most honourable characteristic a creature can have, or rather a high degree of life and rationality, is one which I am at liberty to have views about, but not as a professional. As a matter of fact, my own feeling is . . . '—and here he could go on to tell us his own personal attitude to the question.

The idea of a Sovereign Order of Nature is, in other words, not a purely biological idea: it is an idea originating in biology, but taken over from it and extended to do a fresh, non-biological job. As a matter of biology alone, a creature's place in the scale of nature carries with it no hint of 'authority': to introduce this further element is to elevate the scale of nature out of taxonomy into the realm of scientific mythology. Once this has been done, it has to be recognized that the questions one asks become more than scientific questions—'for all that experiment or observation can show', we said, 'one scientific myth is as good as another'. Here as elsewhere, the attitude which we should adopt towards nature cannot be settled in the way in which one establishes what the facts of nature are; and, when disputes arise about the proper attitude to adopt, scientific considerations alone will be incompetent to resolve them.

In the last 150 years, biologists have given up the idea that creatures can be arranged in a linear sequence. The scale of nature has gone, and in its place we are offered a vast and complex family-tree, showing the genealogical relations (many of them still hypothetical) between the different species of living creatures. As for geology, that has separated off completely, and scientists no longer hope, as even Linnaeus did, to classify inert objects on the same principles as living things. Yet the idea of a Sovereign Order of Nature is not dead. It has roots elsewhere than in the field of biology, and has survived (though transformed) the revolutions produced in biology by Cuvier, Darwin and their successors. From the old idea that the Scale of Nature was fixed, scientists have turned to the idea of evolution; and the theory of organic evolution has, in its turn, been made the starting-point of new ethico-political myths. These are worth a little examination, for they provide as good an illustration as one could find of the difference between scientific theories and scientific myths.

At the present time, the leading advocate of 'Evolutionary Ethics' is Dr Julian Huxley.[4] For him the object of biological study is not only an understanding of how things are: it is also a recognition of how they ought to be. From discoveries about evolution, from 'the new knowledge amassed by biologists during the last hundred years' (he tells us) we can obtain 'definite guidance as to how we should try to plan social and political change'. When we search for the foundation of ethics, we should look to the biologists' theory of evolution, for the object of our search is to be found in the process of evolutionary development: 'the ultimate guarantees for the correctness of our labels of rightness and wrongness are to be sought for among the facts of evolutionary direction'. What Nature is, and how it has reached its present state, he tells us as a

4. His own writings on this subject and those of his grandfather T. H. Huxley are conveniently reprinted together in *Evolution and Ethics 1893 – 1943* (1947): the quotations discussed here are taken chiefly from the Herbert Spencer lecture on 'Evolutionary Ethics'.

biologist; and from there he goes on in his capacity as a prophet, to tell us both where Nature is going and more . . . that it is our job to help her along.

Can we really draw this sharp distinction between the tasks of a biologist and those of a prophet? Dr. Huxley would deny this, and it is my main business to establish, on the contrary, that we can. So with one eye on his text, and the other on my previous argument, let us inquire what sort of connection he is asking us to recognize between the facts and theories of biology and our ethical and political ideals. Does a proper understanding of the theory of evolution dictate to us what our ideals should be? Or is the connection a more tenuous one—does one have to tinker with the results of biological study before they will yield conclusions of this sort? Only in the first case can the biologist speak about ethics in a tone of authority: once he begins to expand or extend biological concepts in ways which are not needed for purposes of explanation, the difficulties we have already encountered will begin to arise—there will no longer be any way of selecting from among all the views that might be put forward one particular 'biologically correct' one.

How then, for Julian Huxley, do the discoveries of biology support ethics? One must assume first that the scientific facts are intended in the ordinary way as *reasons*, and for him conclusive reasons, for selecting certain policies, principles and practices and rejecting others. To this, T. H. Huxley (Julian's own grandfather) would certainly not have agreed. 'Cosmic evolution', he wrote, 'may teach us how the good and evil tendencies in man may have come about; but, in itself, it is incompetent to furnish any better reason why what we call good is preferable to what we call evil than we had before.' Still, this is the first possibility we must examine.

If we interpret Julian Huxley in this way, we can give a summary selection of the backing he offers for his ethical claims in his own words:

> 'Evolution, from cosmic star-dust to human society, is a comprehensive and continuous process. During the process new and more complex levels of organization are progressively attained, and new possibilities are thus opened up to the universal world-stuff. Evolution on the inorganic level operates over an appalling vastness of space. Finally on our

earth the world-stuff arrived at the new type of organization that we call life. During the thousand million years of organic evolution, the degree of organization attained by the highest forms of life increased enormously. Finally there is, in certain types of animals, an increase in consciousness or mind. There is thus one direction within the multifariousness of evolution which we can legitimately call progress. Biological or organic evolution has at its upper end been merged into and largely succeeded by conscious or social evolution. It is only through social evolution that the world-stuff can now realize radically new possibilities. And in so far as the mechanisms of evolution ceases to be blind and automatic and becomes conscious, ethics can be injected into the evolutionary process. The evolving world-stuff can now proceed to some understanding of the cosmos which gave it birth. . . .'

This is heady stuff, and one hesitates before calling it science. Still, up to this point there is nothing in it of an unambiguously ethical sort. Later on, however, he is making undeniably ethical claims:

'Social organization should be planned, not to prevent change, not merely to permit it, but to encourage it. . . . Social morality is seen to include the duty of providing an immense extension of research. . . . Knowledge, love, beauty, selfless morality, and firm purpose are ethically good. . . . The major ethical problem of our time is to achieve global unity for man.'

How does he make the transition? The crucial passage in which the jump from science to ethics is taken is as follows:

'When we look at evolution as a whole, we find, among the many directions which it has taken, one which is characterized by introducing the evolving world-stuff to progressively higher levels of organization and so to new possibilities of being, action, and experience. This direction has culminated in the attainment of a state where the world-stuff (now moulded into human shape) finds that it experiences some of the new possibilities as having value in and for themselves; and further that among these it assigns higher and lower degrees of value, the higher values being those which are more intrinsically or more permanently satisfying, or involve a greater degree of perfection.

'The teleologically-minded would say that this trend embodies evolution's purpose. I do not feel that we should use the word purpose save where we know that a conscious aim is involved; but we can say that this is the *most desirable* direction of evolution, and accordingly that our ethical standards must fit into its dynamic framework. In other words, it is ethically right to aim at whatever will promote the increasingly full realization of increasingly higher values.'

From this passage, one can extract the fundamental ethical principle from which all Dr Huxley's ethical conclusions are presented as following; namely, that 'it is ethically right to aim at whatever will promote the increasingly full realization of those values which are more intrinsically or more permanently satisfying, or involve a greater degree of perfection'. Where does evolution enter into this? It does not come in at all. The reference to this notion, indeed all reference to biology, has dropped out. Regarded as reasons for our choices and preferences, or as premises of a formal argument, all the statements in the preliminary scientific 'build-up' cancel out before the vital step into ethics. (In any case, one might add—though this is by the way—the principle finally reached is too sloppily stated to be of any genuine help to us: of course the 'rightness' of an action in some way 'involves' the 'degree of perfection' of whatever it 'promotes', but how? To say only this much is to stop at the very point where the real business of philosophical ethics begins.)

Our first hypothesis must therefore be dismissed. However the facts about organic evolution are intended by Dr Huxley to support his ethical views, it is not by serving as 'reasons' for them. When he says that 'any standards of rightness or wrongness must in some way be related to the movement of the evolutionary process through time', the connection he sees between morality and evolution is evidently of another sort. Let us see, then, just what he means by the term 'evolution' in this context, and how this compares with the meaning the word has in strictly biological contexts. 'Here as elsewhere', wrote T. H. Huxley, 'names are noise and smoke; the important point is to have a clear and adequate conception of the fact signified by a name.' So is the 'evolution' which for Dr Huxley has a clear connection with ethics simply the biologist's notion? Or is there, for him, something more to it?

First, a word about the part the notion of evolution plays in science proper. This notion was introduced to account for the way in which the distribution of different biological species changes as time goes on: it is concerned, that is, with the relative success of different stocks of creatures in the competition for survival and multiplication. A corollary of this is, that zoologists do not study some single process called 'evolution'; they study an unlimited number of processes, all of which are equally entitled to be called 'evolutionary processes'. The extinction of the mammoths, the growth in the British fulmar population, the development of the lion from its fossil forbears; these are all evolutionary processes, no more and no less than the process by which our own stock sprang from the early primates. It is, in other words, not a unitary process called 'Evolution' that zoologists study, so much as 'the evolution of this-or-that species in a particular environment from such-a-stock'. (There may of course prove to be close similarities between these different processes, which entitle zoologists to explain many of them as instances of a single *type* of process, but that is another matter.) So the whole theory of evolution is such as an intelligent lion or an articulate ant might subscribe to. Although, as men, we may be more curious than a lion or an ant would be about the details of our own remote ancestry, this particular problem is for the zoologist just one among others. If it raises problems of special theoretical interest, well and good: but for many purposes it is more practical to study fruit-flies.

What sort of implications, then, can the theory of organic evolution have for ethics? Its subject matter is the external, or foreign relations between each kind of creature and the others with which it is in competition—between the mammoths, for instance, and the creatures that displaced them. The domestic relations between the individuals of a given species are its concern only in special circumstances, if they react upon the survival-value of the stock—if, for instance, it turns out that civil strife among the mammoths was a factor in their disappearance. But even in that case a biologist could say professionally only that this explains how it was that they became extinct, not that it shows that they had been wicked to quarrel. So the idea of natural selection between species and their subspecies and varieties—for which the biological term 'evolution' is an abbreviation—could have a direct relevance to our ethical problems only if the human race were seriously threatened by the rise of, say, a race of giant ants.

But has not the course of evolution a direction? Is there not as Dr Huxley puts it 'one direction within the multifariousness of evolution which we can legitimately call progress'? And does this not have ethical significance for us? Certainly, it is natural enough that one should put the different sorts of creature there are in an order, according to their degree of evolutionary success: such a classification may be of real use to biology. We can, therefore, understand a biologist's choosing to define 'degree of evolutionary advance' in terms of the qualities associated with survival: namely, ability to control the environment, independence of the environment and capacity to develop further in the same direction. But these criteria, which may serve as a test of 'biological progress', have no more to do with morality than the original notion of natural selection. They, too, are concerned with the actual external rather than the proper internal relations of biological species.

It can hardly, therefore, be the working zoologist's notion of evolution that Julian Huxley has in mind when he talks about the relevance of Evolution to Ethics. Nor does he seriously suggest that it is: Darwin's conclusion that the 'ethical yard-stick' is concerned less with evolution, in the sense of natural selection, than with a man's relation to his fellows, Huxley considers a feeble one. For him, 'evolution' means more than natural selection. It is a composite notion and a more grandiose one for which the biological term serves only as a starting-point. Here is the recipe for producing it. First, select from the countless evolutionary processes that biologists study the particular one which has led up to the appearance of man—this puts man in a preferential position, whose justice the lion or the ant might question, but the justification would be man's greater degree of 'evolutionary advance'. Then tack on to the beginning of this historical sequence a series of physico-chemical events leading up to the appearance of the first living creatures, and at the latter end treat the development of civilization and technology as a continuation of the biological trend 'by other means'—in this way, as Huxley puts it, we 'extend the concept of evolution both backward into the inorganic and forward into the human domain'. Finally, christen your conceptual artefact 'the cosmic process' and present it as a golden thread leading from the remotest past up to the present day and on into the future. This, not 'evolution' in the pure biologist's sense, is the touchstone Julian Huxley offers for solving our ethical problems.

Can the composite notion constructed to this recipe claim to be a genuine scientific notion? It could do, if the extended use of the term were forced on us by the facts—that is, if we needed to appeal to it in order to explain the occurrence of events which must otherwise remain unexplained. But the motive for extending the meaning of the word 'evolution' as Huxley encourages us to do, is not to explain anything fresh: it is something very different.

We find here both of the features characteristic of a scientific myth. The key-term used has to be understood, not in its straight scientific sense, but in an extended one; and this extension is made for other than scientific motives—so that, when differences of opinion arise, there is, scientifically speaking, no longer any way of deciding between them. We are to use the term 'evolution' to cover not only biological processes, but also certain inorganic phenomena which took place before the appearance of living creatures on the earth and, in addition, the development of human society since the emergence of man. This means giving the term 'evolution' a very different shape from its zoological one, and forcing it into quite a fresh jig-saw. The claim that biology is behind Huxley's ethical claims cannot therefore be allowed; and once we start changing the meaning of the term 'evolution', to put it to non-scientific tasks, biology can no longer resolve our difficulties when incompatible views are put forward.

There are in any case ambiguities about the recipe: the zoological sequence of events leading up to the appearance of man may be determinate enough, but which exactly are 'the' inorganic processes leading up to the appearance of the first living organisms?

How are we to select these favoured processes from all those taking place 'over the appalling vastness of space' a million years before the first cells formed? Even worse, what criteria are we to use to select 'the' process of social development, except by begging the very ethical questions that Evolution was going to help us to solve? The biological criteria of 'degree of evolutionary advance' are hardly enough, for emphasis on 'control over and independence of environment' suggests that technological skill is the sole criterion of social advance. Yet if we add further criteria and judge social progress on other grounds, we cannot help being influenced by our own ethical preferences when we make our selection.

Once this stage is reached, the impossibility of selecting between ethical views on biological grounds becomes absolute. For now, so far from our finding the ultimate guarantees for the correctness of our labels of rightness and wrongness among 'the facts of evolutionary development', the present-day direction of that development becomes itself a matter of opinion. The question 'Which way is Evolution moving to-day?' now becomes equivalent to the question, 'Which way *ought* society to develop to-day?': the appearance of biological fact simply conceals the ethical character of the point at issue. Restoring horse and cart to their proper positions, we now have to look for the ultimate guarantees of our generalized labels of 'evolution' and 'evolutionary progress' among the facts of morality. Having extended the notion of evolution into the moral sphere we can no longer decide unambiguously what Evolution is, unless we have already *settled* our ethical disagreements.

Again, as we foresaw, when different views about the relevance of evolution to ethics are put forward by other writers, there proves scientifically to be no way of choosing between their views and Julian Huxley's. He is certainly not the only man who invites us to think of the history of the universe in this way: we all know of others who represent the development of the cosmos as, so to speak, the progress of a cosmic band-waggon—a historical juggernaut sweeping irresistibly out of the past and on into the future. It is not surprising to find that Hegelian and Marxist theorists, too, include biological evolution as one phase in their 'dialectical process'. But whichever account of the 'cosmic process' we look at, the same difficulties arise the moment one tries to draw concrete morals from it. For instance, ought we to jump on to the waggon and travel with it, or ought we to try to stop it? And if we ought to jump on to it, what is its destination? In each case there are several opinions, and even those who advocate jumping on the waggon disagree about its route. Hegel thought the King of Prussia was at the wheel, but this view is now implausible; nowadays dialecticians do not always agree whether the main line goes through Moscow or through Belgrade; while Julian Huxley would prefer to think the waggon called at the United Nations Headquarters. But how can we settle which of them is right?

There are also those who would have us put the waggon into reverse. T. H. Huxley himself is a good example—'The ethical progress of society depends,' he wrote, 'not on imitating the cosmic process, still less on running away from it, but on combating it.' Between him and Julian there is therefore a flat contradiction. Can we choose between their views? Not on scientific grounds; for the question, whether social and moral progress are to be regarded as a continuation of the process of natural selection or as a reaction against it, is not one which can be answered by appeal to observations or experiments. There is no question of phenomena turning up one day which could be accounted for only on one or the other of the two views, as would have to be the case if the issue were really a scientific one.

What is the real point at issue, then? If we ask this, we shall see the nature of Evolution Myths more clearly. When one considers why each of the two Huxleys chooses to 'extend' the notion of evolution as he does, the most one can find is a difference of temper between them; there is no disagreement about the facts. One of them sees social development as a continuation of the process of natural selection, the other as a reversal of it; and these different attitudes to 'evolution' reflect the different things they want to do with the notion.

For Thomas Henry Huxley, 'evolution' represented a Challenge: for Julian it is a Talisman. T. H. was concerned to point out important differences between the methods of natural selection and of social and moral development. What struck him about natural selection was the starvation and slaughter through which species supplant one another, and especially the fate of the weak and the meek. This brutality, he felt, was just the thing which, in our relations with our fellow human beings, we must fight against. He therefore painted a picture of Man as pitted against Nature, and turned his spotlight on to the competitive aspect of biological evolution, contrasting the brutality and destructiveness of natural selection with the harmony desirable in society and by implication criticizing a political economy of unrestricted competition.

'The struggle for existence tends to eliminate those less fitted to adapt themselves to the circumstances of their existence. The strongest, the most self-assertive, tend to tread down the weaker. But the influence of the cosmic process on the evolution of society is the greater the more rudimentary its civilization. Social progress means a checking of the cosmic process at every step and the substitution for it of another, which may be called the ethical process; the end of which is not the survival of those which may happen to be the fittest, in respect of the whole of the conditions which obtain, but of those who are ethically the best.'

T. H. Huxley, in other words, used the devilish mechanism of natural selection as material for a parable—'Nature red in tooth and claw' was to do duty as a symbol of our own competitive brutality. For these purposes, one must agree, the 'cosmic process' was well fitted to play the part of the Devil.

When Julian Huxley writes about his 'Evolutionary Ethics', however, he is not interested in producing a parable directed against cruelty and selfishness. So far from spoiling for a fight with nature, he feels it most important to have Nature behind him. T. H. Huxley saw in the cosmic process an enemy worth fighting, and set his great bearded jaw against it: Julian turns to the same process for reassurance. The connection he sees between evolution and ethics is not, after all, a logical or scientific one. What he is looking for, he says, is 'reassurance' in the face of a 'hatefully imperfect world'; a 'cosmic sanction for ethics', a 'natural foundation on which our human superstructure of right and wrong may safely rest'—in a phrase, *ein' feste Burg*. His aim being different, he casts the cosmic process in a different role: we are to think of Evolution not as the Devil but rather as the Deity.

Once this is recognized, a great deal becomes clear which was formerly mysterious. The tone in which he writes, for instance, is so unlike the cold, scrupulous tone of a learned scientific journal:

> 'Compared with what a protozoan or a polyp can show, the complexity of later forms of life, like bee or swallow or antelope, is stupendous, their capacity for self-regulation almost miraculous, their experience so much richer and more varied as to be different in kind.'

This is the sort of writing one expects rather in Paley's *Natural Theology*. Indeed, whenever Dr Huxley writes about this topic, one keeps coming across passages written in the language of wonder, not of theory—of religion, not of science—passages designed (in Pascal's phrase) 'to enkindle, not to instruct'.

To do Dr Huxley justice, he has no serious intention of disguising the theological character of his writing, for his account of Evolution is openly presented as the theology of his own 'Religion without Revelation'. It is only because he is a scientist that one at first expects his writings to belong to popular science, rather than theology. Yet this is a mistake which could really have been made only in the twentieth century: until our own time, scientists have always been accustomed to writing on natural theology, and to using the scientific discoveries they described for theological ends. Evolution, then, takes on for Dr Huxley most of the jobs of the discredited Deity:

> 'The so-called immutable laws and will of God, which are invoked to guarantee the principles of ethics, turn out to have been extremely changeable; and the principles of ethics have changed with them.'

God has failed: we must therefore put our trust in Evolution, and vary our ethical beliefs as it directs. Evolution now becomes not only the Source of Comfort and Reassurance—'evolutionary ethics is of necessity a hopeful ethics'; it figures also as the Immanent and Omnipresent Creator, that 'comprehensive and continuous process' which 'moulded the world-stuff into human shape', and whose Agent it is man's privilege to be—'Man is not only the heir of the past and the victim of the present: he is also the agent through whom evolution may unfold its further possibilities.' All the wonders which for Archdeacon Paley were evidences of the existence of God can on this view be put to the credit of Evolution. As for us, we can be confident that the direction in which it is going is the right one; so, if only we mount the waggon, all will be for the best.

Now it is not my purpose to criticize these views from the point of view of theology: I am content with showing that this is what they are. Yet there are several drawbacks to a religion of Evolution which even a layman will feel. To begin with, it must be a religion of limited appeal, seeing that the Myths on which it is based are expressed in a way properly intelligible only to professional biologists—for others the analogy between moral progress and biological progress will have no great force. Again, the Supreme Being it presents to us is hardly an object of respect: Dr Huxley himself calls it 'a glorious paradox' that 'this purposeless mechanism, after a thousand million years of its blind and automatic operations, has finally generated purpose' now that 'Man the conscious microcosm has been thrown up by the blind and automatic forces of the unconscious macrocosm'. Worst of all, perhaps, it is not a religion with much power to stimulate human endeavour; for those who have not sufficient scientific training to recognize the limited scope of the analogy between moral and biological development (that is to say, for most of us) its effect would probably be just the opposite. This was certainly the effect of its Hellenistic counterpart, the religion of $T\acute{v}\chi\eta$, the personification of chance, which had a vogue after the Olympian Gods had become discredited.

All that matters for my present argument is the logical point, that the two Huxleys' conclusions about the relevance of biological evolution to ethics are not themselves scientific conclusions. Logically speaking, there is no reason why you should not hold either position—or both. There is a good deal in what they both say, and which aspect of the theory of evolution you could consider the more worth stressing will depend on your immediate aim and interests. This aim may be partly a matter of temperament: one could write an essay contrasting the attitudes of T. H. Huxley's two author-grandsons—J. S. cheerfully accepting biology and progress, Aldous critical of both. (The character of Shearwater in *Antic Hay* would make a natural starting-point.) Partly it will be a matter of historical circumstances: we in the mid-twentieth century have had enough brutality to contend with nearer home for us not to worry, like T. H. Huxley, about the brutality of natural selection, while, conversely, there would have seemed less need sixty years ago for the kind of comfort that the Cosmic Process affords Julian Huxley now. Whichever position you choose, one thing alone is essential: not to suppose that there is any material conflict between the positions or conclude, as Julian Huxley does, that 'T. H. Huxley's antithesis between ethics and evolution was false'. In this context, the word 'false' is far too strong. What we are called on to decide is not which of two scientific theories is better in accord with the facts of observation: the task is now to decide in which of two Scientific Myths we find the more congenial attitude to Nature.

Several questions about Julian Huxley's account of the matter are, however, still unanswered. He is evidently reassured by the thought that the 'cosmic process' is at hand to serve as a 'foundation' for ethics. Yet how does this come about? Supposing ethics to need a prop, how could evolution serve as one? It is not clear in what sense the one can be said to 'support' the other, and the sceptical might ask whether evolution provided a genuine foundation for ethics any more than a row of whisky-drinkers genuinely propped up a bar.

Of course, the intellectual support of science is nowadays worth claiming for any position, however tenuous your justification. One remembers how the impressionist painters, too, claimed that their techniques were more scientific than those of their predecessors had been. If olive and brown and purple were not in the spectrum, they argued, painters should not allow them on their canvases, and only pure colours, the colours of the spectrum, should be used. Yet, however admirable the results of the impressionists' decision, however fruitful the technical experiments they were inspired by their reading of physical optics to make, the idea that the discoveries of physics could *justify* their techniques was an illusion. A painter may *choose* to use only the colours of the spectrum, but it will be by his works, not by science, that his choice will be justified. Whether the impressionists succeeded in their aim only a beholder is in a position to say—the purity of their colour is a matter, not for the spectroscope, but for the eye.

Is it in this sense that we are to think of evolution as justifying ethics? Is the motive behind the suggestion simply a wish to see some connection, however far-fetched, between the workings of nature and the principles of morals? There is, I think, more to be said. In some ways, the myth of Evolution is like the Atlas myth: both, to stress the obvious, proffer support in contexts in which no support is needed. But both also have deeper motives which are worth uncovering. Atlas, for instance, is often thought of as show-ing only how ignorant the Ancients were: had they known a little more about the solar system, we feel, they would have seen that there was no need for an Atlas, for he was the answer to the question 'What holds the Earth up?', a question which need never arise. Yet surely Atlas was the product not merely of ignorance. There were a vast number of things besides the mechanism of the solar system of which the Ancients were ignorant; but very few of them gave rise to myths. Only where this ignorance was of impor-tance, where it seemed to mean insecurity, was a myth born. Whether or no we have a clear and satisfactory picture of the solar system may not matter to us directly, for after all there are few ways in which we make direct use of this understanding. It is, however, by their indirect effects that myths get hold on us, and in this case the effects are considerable, so that one is not surprised to find that Atlas has counterparts in many mythologies. The stability of the Earth becomes a symbol for so much else. If we have no assurance of that, what else can we trust? No better outlet could be found for 'an anxious fear of future events'; there could be no guarantees, no reassurance about the future, without the confidence that the ground below our feet rested on good, strong shoulders.

Similar motives are probably at work in generating Evolution myths. The support given by Evolution to ethics serves as a source of confidence in our moral ideas, rather than as an intellectual justification of them. In a time of uncertainty and change it is natural for us to wonder whether, in doing what we think right, we may not simply be wasting our time; as a result, we may feel the need of some assurance that there is (so to speak) some future in ethics. People become unhappy about the prospects of virtue paying any dividends, and begin to look elsewhere for a security. This may, intellectually, be as much the result of a misconception as the demand for an Atlas to hold up the Earth—I think myself it is. Still, quite apart from intellectual questions, there is a further point. If you can paint a picture of social and moral development as being all of a piece with biological development, this may help you to feel that morality is something long-standing and of proved worth, something with roots in the universe and no mere human makeshift. Then next time you feel the fear that morality may after all not prove a paying proposition, you can at least (on this view) comfort yourself with the thought that such an ancient institution as the cosmic process is not likely to default.

An anxious fear that the Earth itself may be insecure: that is what gives the Atlas myth its strength. The same kinds of motives can be seen behind many of the old stories. Fear of the sea and the storm, fear of the harvest failing, and the hope of averting these calamities by propitiation: these, not mere ignorance, were responsible for Poseidon, Wotan and Ceres. The same motives remain strong, even though our ignorance may be less. An anxious fear of the remote and unknown past, and of the remote and even more unknown future: these lead us to look for eschatological morals even where there is no hope of finding them—in physical cosmology. Again, the desire that morality should unquestionably be worth while, that the importance to us of our own affairs should find a reflection through all the history of the universe: is it fanciful to see this as the motive behind the Myths of Evolution?

My most serious doubt is whether biological prose, however highly coloured, could ever be an adequate medium for putting this desire into words. It is perhaps inevitably a poetical desire, this impulse to read our purposes into the world of nature. To find the essence of virtue, not in the day-to-day give-and-take of domestic life, but embedded somewhere in the impersonal structure of things: that would be to discover a true Talisman; and the writers who have captured the impulse most successfully have done so in a prose which was very near to poetry. Here is how Virginia Woolf put it, in *To the Lighthouse*:

'As summer neared, as the evenings lengthened, there came to the wakeful, the hopeful, walking the beach, stirring the pool, imaginations of the strangest kind—of flesh turned to atoms which drove before the wind, of stars flashing in their hearts, of cliff, sea, cloud and sky brought purposely together to assemble outwardly the scattered vision within. In those mirrors, the minds of men, in those pools of uneasy water, in which clouds for ever turn and shadows form, dreams persisted, and it was impossible to resist the strange intimation which every gull, flower, tree, man and woman, and the white earth itself seemed to declare (but if questioned at once to withdraw) that good triumphs, happiness prevails, order rules; or to resist the extraordinary stimulus to range hither and thither in search of some absolute good, some crystal of intensity, remote from the known pleasures and familiar virtues, something alien to the processes of domestic life, single, hard, bright, like a diamond in the sand, which would render the possessor secure.'

Science and Our View of the World

The ambassadors of the intellect in high places often express their regret at the 'divorce' between natural science on the one hand and philosophy on the other. In it they see not just an intellectual lacuna—the sort of thing which one would like to see filled, for reasons of disinterested curiosity alone, like the blanks in an all-but-completed crossword puzzle. To them the divorce appears a sign, a symptom, perhaps even the cause of greater evils and more radical distresses. What the scientist and the philosopher should aim at, they suggest, is a reunion of their disciplines: a 'synthesis' is called for of the results of the special sciences, and this synthesis is to provide a more comprehensive 'world-view' than can be obtained from any one of the special sciences alone.

The response of working scientists and philosophers to this request must be a disappointment. Attempts at reconciliation lead for the most part to mutual suspicion, and it becomes clear that the re-marriage cannot hope to be a love-match. To most scientists 'the acquisition of knowledge about the world of experience' seems sufficient intellectual exercise: this, they declare, is a field 'wide, rich enough in changing hues and patterns to allure us to explore it in all directions'—the 'dry tracts of metaphysics' beyond they willingly leave to others.[1] It is true that some scientists, a few, do set out on the quest for a synthesis, but to their colleagues these activities are, more often than not, an embarrassment.

1. Max Born, *Atomic Physics* (Blackie 1937), p. 258.

The indifference is requited. In the years which have passed since G. E. Moore attacked Herbert Spencer's views on ethics in *Principia Ethica*, professional philosophers (in England at any rate) have hardly felt that philosophy in Spencer's manner—synthetic philosophy, to use his own phrase—was worth serious attention. They have of course felt bound to review, as they appeared, the philosophical writings of Spencer's successors, men such as Huxley, Waddington, Eddington and Jeans; but they have had no difficulty in detecting in all their works enough linguistic confusion and logical sloppiness to dispose of them, at any rate to the satisfaction of the reviewers.

Yet, though spurned by the comrades on either side, the bridge-builders themselves have not felt disposed of. They refuse to agree that Moore's devastating broadsides have found their mark: all they will concede is slight damage to some outlying bastion. Here is Dr Waddington, for instance, on 'the melancholy fate of Herbert Spencer':

> 'Poor Spencer! He was cajoled or bamboozled by literary men into behaving as though he was talking not about phenomena but about forms of words. Rather half-heartedly he stated, or implied, that what he meant by "good" was "productive of pleasure"; and the critics (e.g. Moore) showed their gratitude at not being asked to raise their eyes from their books by pointing out that that would not do at all.'[2]

The situation is a peculiar one. It is not a question, as one might at first suppose, of whether this activity of 'bridge-building' is being well or ill performed: the question is rather, whether or no the activity gets one anywhere, what it is directed towards, whether it is more than an elaborate intellectual game. Evidently the synthetic philosophers themselves have no doubts: they feel that their activity is both legitimate and important, and the standard lines of philosophical criticism, they claim, are simply irrelevant. If this claim is just, the consequences are important: the professional philosophers have missed the point, and have failed to recognize

2. C. H. Waddington, *Science and Ethics* (1942), pp. 136–7.

what the bridge-builders are about. Why, then, have the synthetic philosophers felt so unfairly treated? And why have the professional philosophers, for their part, been so contemptuous of 'bridge-building'? The material we have been looking at in this essay will help us to answer these questions, and to see more clearly the scope and limitations of any world-view synthesized from the conceptions of the natural sciences.

'It is not words the synthetic philosopher is interested in', we are told: 'it is *phenomena*. He is not concerned to frame definitions but rather to see the world aright. Success for him would be represented not by an impeccable analysis of the term ''good'' but by a proper picture of the order of nature and of man's place in it.' This of course is an entirely laudable programme and, if philosophers had not so much evidence of the harm done in the course of such inquiries by unwittingly accepting bizarre definitions of 'good', there would be nothing to question in the statement of it. It is when we leave programmes and try to get down to business that trouble starts.

What, for instance, is our test of a 'proper' picture of things? Is there a single criterion of 'propriety' capable of general application? Or will the criterion depend on an author's precise intentions, and vary with them? If one is building up a world-view from scientific bricks, we must expect the last to be the case. Synthetic philosophers do not all have the same aims. It is not the hypothetico-deductive explanation or mathematical representation of phenomena which is their aim, nor are our expectations their target. They wish rather to alter our attitudes to all sorts of things, and each author has in this respect his own preoccupations. Even when they seem to be discussing the same subject, such as 'the connection between evolution and ethics', their arguments may (as we saw) have completely disparate intentions; and in consequence there is no question of a final decision between them, no definite test—in the laboratory or elsewhere—on the basis of which one view can be established and an ostensibly opposed view ruled out. That is what makes it so inappropriate to talk in this context of truth and falsity, in any black and white kind of way. When synthetic philosophy is under discussion, we have instead to talk in terms of 'seeing the point of' one doctrine, and 'recognizing that there is something in' or 'being in sympathy with' another. In this field, black-and-white logical judgments have no place.

These conclusions apply with particular force to the world-views which are sometimes produced as 'following from' the different branches of science. If there is no final choice to be made between different views about the relation between ethics and evolution, how much less can we talk about a genuine contradiction between (for example) what people call the 'physical' and 'biological' pictures of the universe.

It is true that, if you confine your attention to the materials of one single branch of science at a time, you will be able to construct from the different sciences world-views, or views of Nature, so completely opposed in temper that they seem impossible to reconcile. The optimistic picture of evolution as a steadily-advancing process, and the pessimistic one of man as engaged in a rearguard action to stem the rising ride of entropy: it is no wonder that Mr John Heath-Stubbs cannot help feeling that these 'must seem to be contradictory, at any rate to our limited intelligence'.[3] None the less, they are not so much contradictory as complementary: to treat the contrast between them as anything stronger is to misunderstand the relation between the expressions of these visions of Nature and the 'scientific evidence' offered in their support. Do the physicist's discoveries justify gloom, and the biologist's hope? This seems to be the implication of the term 'contradictory', and it is a false one. For, if to the outsider there appears to be an air of burgeoning optimism about the science of life which that lifeless subject, physics, so notably lacks, the reason does not lie in anything that physicists and biologists have *discovered*. What is relevant here is, rather, the distinction we ourselves have drawn between 'physics' on the one hand and 'biology' on the other—the manner in which we have come to sort out the subject-matter and methods to be labelled 'physical' from those which are to be called 'biological'. This preliminary act of ours, not anything scientists have found out subsequently, is what makes biology such a fruitful source of imagery for the optimists, and physics an inexhaustible well for the pessimist.

3. *Poems of Leopardi*, tr. Heath-Stubbs (Lehmann 1946), Introduction, p. 13.

The point is worth following up a little way; so consider how we do make this selection, and in particular how we tell an 'organism' when we see one. The most striking and characteristic feature of organisms, the thing which marks them off from 'inert' matter, is their activity. Teleological words, words referring to 'conduct', apply only to living things, and in their literal and original senses do not make sense of the inert. What goes on under the skin may be interesting and important but it is, from the logical point of view, accidental. Take Mrs Jones' dog Fido and change, as much as you please, his bones and muscles, brain and stomach: provided he still wags his tail for her in the old enchanting way, comes to heel when his mistress calls, eats the joint and barks at Scotsmen, she will feel that she can justly say 'That's the same old Fido'.

The biologist, it is true, soon becomes as interested in the structure of organisms as in their activity. Teleological accounts of their behaviour he comes to regard as a second-best, for he would like to be able to connect up everything in an animal's conduct with his knowledge of its internal structure: in this way it ceases to be Fido and his ways so much as the concentration of phosphorus in his nerves that becomes important. None the less, it is still organisms in whose structure the biologist is interested; and it remains the activity of organisms which is their chief defining characteristic. By 'signs of life' we mean not structures, for these are shared with fossils and the dead, but activities; and however mechanistic a biologist may be in his methods (and very properly so) it is still the structure of *living* matter whose mode of functioning he studies. His subject-matter, in other words, is still pre-selected as coming from the sorts of things which can meaningfully be said to do, want, hunt for, attack and eat things. By contrast, the subject-matter from which the physicist starts consists of the entirely inert, i.e. of those things to which teleological words are least of all applicable.

It is this division of the sciences, according to their methods, the questions they ask and the subject-matter they study, which lies behind the notion that physics and biology imply opposed 'world-views'. There is no need for physicists or biologists to make any discoveries about the world in order for this opposition to be established: it was built in to the terms 'physical' and 'biological' in the first place. Nor does the division correspond to any necessary division of scientists by temperament: not all biologists are optimists, nor are all physicists gloomy. For many scientific purposes it is irrelevant whether a specimen is living or lifeless—these are the ones we have chosen to call 'physical'; but if, when a physicist weighs eggs, he does not ask whether they are fertile or addled, it does not follow that he takes them all to be addled. So biologists and physicists have not come respectively to the conclusions that the world is a cheerful or a gloomy place: it is we who have beforehand allotted them their respective fields of study.

At the same time, though the idea of physics and biology 'implying' opposed world-views may be a misunderstanding, there is another, more authentic contrast between the sciences. The poet or philosopher who, like Leopardi, sees Man as separate from Nature will find in physics the imagery to express his vision of things since (simply because physics is physics) it contains no way of referring to the desires, aspirations and aims of men. To introduce references to these things would be to leave the field of physics—by definition, so to say. The physicist's world, then, is to the poet's eye a 'dead' world, a world characterized by 'the destructive potency latent in the volcano, and the vast empty spaces through which wander the lifeless and purposeless stars'.[4]

It becomes clear, however, that physics is being used merely as a source of imagery if we notice the queerness, in this context, of calling the stars "lifeless and purposeless'. The man who walks so happily along the cliff-top slips and falls to the beach below: the doctor examines his body and pronounces it 'lifeless'. The craftsman, skilled through years of practice and absorbed in his work, loses his job in an economic upheaval: he draws his dole and wanders round the streets 'purposeless'. If the words 'lifeless' and 'purposeless' have the meanings and the (depressing) associations they do, it is because they have acquired them in just such contexts as these. Those things are, in this sense, lifeless, which might have been alive but are not; and those purposeless which might be intent on a fruitful goal, but are not. As for the stars, it is not in this sense that they are lifeless and purposeless: they no more resemble a man

---

4. Heath-Stubbs, *loc. cit.*

who has just lost his life or his job than they do a living and busy one. They are, one might more properly say, neither living nor lifeless and neither purposeful nor purposeless; since there seems to us nowadays no way of applying to such things as the stars either of these two oppositions.

Have we, then, any cause to lament? Surely not: if astronomers were, on the other hand, to start treating the heavenly bodies as living creatures once more, we should then begin to have some reason for worrying. Not the least of the merits which Edmund Halley saw in Newton's theory was its power to banish anxieties of this sort:

> 'Now we know
> The sharply veering ways of comets, once
> A source of dread, nor longer do we quail
> Beneath appearances of bearded stars.'[5]

The inertness of the stars—the only sort of 'lifelessness' that can be attributed to them—is surely preferable to their possible malignancy: to lament about it is simply to project one's own despair on to the skies.

As the subject-matter of physics provides the imagery of despair, so that of biology provides the imagery of hope. The biologist's business is with the living, with things that are pursuing many and achieving some ends, desiring much and fulfilling some of their desires; and this preoccupation he shares with the optimistic poets and philosophers, who think of Man, not as standing apart from, and pitted against Nature, but as one with all her works. But again this does not imply that for the biologist everything is going up and up and on and on!

The logical relation between the arguments presented in support of the 'physical' and 'biological' visions of Nature and the visions themselves is, therefore, the reverse of what it at first sight appears to be. If the doctrines that 'we are powerless to stop the Universe running down' and that 'we are an integral part of the creative process of Evolution' were straight scientific theories, the arguments supporting them would, logically speaking, be reasons for

5. Newton, *Principia* (Cajori edition), p. xiv.

accepting them. The opposition in temper would be accompanied by oppositions of fact. This however is not the situation. The evidence and arguments brought forward in support of one doctrine or the other are not what justifies us in accepting or rejecting it: rather it is the temper of each doctrine which determines what selection of 'evidence' will be presented or spotlighted. Whether you are to accept either doctrine, either vision of Nature, either set of arguments, depends not on the quality of the reasoning, but on how your own attitude to the world already compares with that of the author.

If you are a political optimist, you may be attracted by such lyrical passages as this of Dr Needham's:

> 'The new world order of social justice and comradeship, the rational and classless world state, is no wild idealistic dream, but a logical extrapolation from the whole course of evolution, having no less authority than that behind it, and therefore of all faiths the most rational. . . . Even so gigantic a setback [as World War II] cannot shake a faith which is based on the considerations which convinced Drummond and Spencer, Engels and Marx. The way may be long and we may not live to see, but the triumph of the rational spiritual man is sure.'[6]

If however you are a congenital pessimist, you may feel more sympathy for Ostwald's insistence (cited in part earlier in this essay) that

> 'We must in all circumstances learn to accept the fact that at some indefinite but far-off time our civilization is doomed to go under; that the final purpose of human effort lies in Man himself alone; that it is directed towards his transitory existence; and that, in the longest run, the sum of all human endeavor has no recognizable significance.'[7]

But in neither case will you be presenting scientific conclusions, and there is no question of producing experiments which will

6. J. S. Needham, *Time, the Refreshing River* (Allen & U. 1943), p. 41.
7. F. W. Ostwald, *Die Philosophie der Werte*, p. 98.

finally establish the correctness of one view of Nature and finally torpedo the other.

The fundamental difficulty about 'syntheses' of the sciences is, in the last resort, the difficulty about all scientific myths. So long as scientific concepts and doctrines are employed to explain and represent natural phenomena, we can keep some sort of logical control over what is said: arguments and conclusions are open to criticism and verification—and, as Dr Popper is probably right to emphasize, *falsification*. But when we use terms of a scientific origin in an extended manner, as the vehicles of some more-than-scientific attitude to the world, science is neutral between all conclusions. Even H. G. Wells' characters, under the imminent threat of annihilation by a comet, had a choice of attitudes—in similar circumstances, we ourselves would no doubt go, some to the pub, some to the church, and some to continue digging our gardens, and no one of these reactions would be either more or less 'scientific' than the other two.

So perhaps the unwillingness of working philosophers and scientists to consider a re-marriage of their disciplines is, after all, justifiable. Perhaps the demand for a synthesis is, to say the least, premature, and perhaps 'visions of Nature', as opposed to theories about the workings of natural things, are best left to the poets. Certainly Dr Needham cannot compare, as a spokesman for optimism, with Wordsworth; and the gloomy chemist, Professor Ostwald, can do no more than echo the despairing lament of Leopardi on the barren slopes of Vesuvius:

> 'Gaze and see
> How loving Nature cares
> For our poor human race, and learn to value
> At a just estimate the strength of Man,
> Whom the harsh Nurse, even when he fears it least,
> With a slight motion does in part destroy,
> And may, with one scarce less
> Slight than the last, without a moment's warning
> Wholly annihilate.'[8]

---

8. *Poems of Leopardi* (tr. Heath-Stubbs), p. 58: from 'The Broom'.

Conclusions

What is the moral of this essay? It is, I think, no more than this: that we should beware of feeling that scientists are (as it were) initiates, like priests; and also of contrasting the 'scientist' with the 'ordinary man' in a way in which we should never dream of contrasting the tinker or the bus-conductor with the ordinary man. For this habit is liable to weaken our critical faculty, our sense of relevance, and lead us to place too much weight on the *obiter dicta* of scientists. We should soon notice if a tinker or a bus-conductor started laying down the law about things on which his calling did not make him an authority: it is as well to bear in mind that a scientist off duty is as much an 'ordinary man' as a tinker or a bus-conductor off duty.

Once upon a time, perhaps, it would have been less easy to draw a satisfactory distinction between the explanatory and the mythological uses of our concepts. For the Ancients, there was no clear line between Atlas the astronomical notion and Atlas the mythological hero. Still, times have changed, and there is less reason nowadays for overlooking the distinctions I have been emphasizing. Like many other crucial scientific issues, the matter came to a head over the choice between the Ptolemaic and the Copernican cosmology. Until men were confronted with this choice, their own preoccupation with the Earth was reflected in the position which the Earth occupied in physical theory—the centre—so that it was possible, so to speak, to think of the importance for Man of Man's affairs as written in the skies. The result was that, when the Sun instead of the Earth was proposed as the centre of the astronomer's 'system of the world', the new theory aroused more than astronomical objections. It aroused also fear; for the picture of the Earth as being at the centre of things was, as we can now see, not only a theory but also a myth.

To begin with, the new theory was treated in a similar way. It was not felt to be enough that the mathematics of the new system were tidier than the old had been: reasons had to be found for thinking that the Sun was a *worthy* centre of things and its being the source of light and heat was argued, even by such men as Kepler, as showing that the Sun rather than the Earth was the true House of God.[1] In time, of course, people came to see that astronomical questions and questions about the ultimate importance of our mundane affairs were independent; so that their feelings of security and dignity need not remain forever tied up with problems in astronomical theory. When this happened their fear evaporated and they were happy to accept the new Copernican picture. But until it did, the myth and the theory were not clearly distinguished, and the quest for knowledge remained entangled with the quest for security.

Here, as elsewhere, we have come to distinguish between the natural sciences and other disciplines, and to disentangle from the undifferentiated skein in which they first present themselves to us the problems belonging to each. But distinctions which needed making can be forgotten again, and these particular ones are still not always clearly respected. Things which once were fused can be again confused. When we begin to look to the scientist for a tidy, a simple, and especially an all-purpose picture of the world; when we treat his tentative and carefully-qualified conclusions as universal certainties; or when we inflate some discovery having a definite, bounded scope into the Mainspring of the Universe, and try to read in the scientist's palm the solutions of difficult problems in other fields—ethics, aesthetics, politics or philosophy; then we are asking of him things he is in no position to give, and converting his conceptions into myths.

Yet how often we are liable to do this. A scientist broadcasts about his work, for instance, Fred Hoyle or Professor J. Z. Young, and in listening to him we are always ready to find in what he tells us something more than science. What in practice we find particu-

---

1. Cf. S. Mason, *A History of the Sciences*, p. 144.

larly exciting, or disturbing, is not the bits of genuine science he tells us, though these are intriguing enough, so much as the philosophical or theological implications we read into them. Hoyle uses the phrase 'continuous creation' and this seizes our attention, just because it has about it a strong flavour of the Book of Genesis. Similarly, when we hear J. Z. Young talking about those 15,000,000,000 brain-cells of ours: however unjustly, we are liable to understand him, not only as suggesting what is possibly the way of working of objects whose mechanism has up to now been completely mysterious, but also as *explaining away* all kinds of things that we have hitherto had a reasoned faith in, about our minds and about our ideals.

All the same, there is no need to be bullied, or muddled, into accepting as authoritative and established scientific truths conclusions on which the scientist is in no better position to speak than anyone else. Is it suggested that perhaps our ideals are solely a question of brain-mechanism? Well, suppose that identifiable cell-structures or electronic processes were always found to be present in a man's brain if he believed, say, in freedom of speech: that would be a fascinating discovery. But even if it happened, what would it show? Need such a discovery be taken as proving anything about the importance or unimportance of our ideals? Surely not: it is not simply too soon to ask the brain physiologist questions of that sort—this is not the kind of thing he could ever be in a position to pontificate about. Brain physiology is one of the sciences which is likely to make great strides before long, but it cannot prove everything; so do not let us turn to it for guidance in problems to which it could not be relevant.

Fred Hoyle, again, is said to have composed the concluding, unscientific postscript to his book because he had been amazed at the comfort which the devout had been drawing from his first few talks. Now it was legitimate enough for him to argue that such listeners must have misunderstood his talks, and that his astrophysics could not properly be taken as bolstering up their faiths. But he seems to have felt something more than this: that, if properly understood, his theories should have been a source of positive *dis*comfort to religious people. Does this not suggest that he was deceived in the same way as they? For what is puzzling is not people's taking comfort from an astronomical theory—seeing it as a prop for their faith—rather than having religious doubts aroused, and so feeling upset by it. What we *should* boggle at is the idea that either reaction is called for, and that any direct connection exists one way or the other between Hoyle's physics and the attitude we should adopt towards the world.

Professor Herbert Butterfield has made the point very clearly:

> 'When I am engaged upon a geometrical problem, or set myself to study the parts of an intricate machine, there is no reason why my mind should not try to be clear of all affections. . . . When I am thinking about man's nature and destiny, his place in the scheme of time, the posture he should adopt in the universe, and the ends that ought to be his in life, I cannot disentangle myself from my affections. Then, we make not an assertion about anything, but a decision about ourselves—we decide from what angle we will meet the stream of events and what shall be our posture as human beings under the sun.'[2]

Mutatis mutandis, Butterfield's contrast between geometry (or mechanics) and our world-views can be applied to natural science generally and, in its wider application, this contrast has been the central theme of my essay. Throughout the history of science, one finds two threads running side by side and often entangled. Much of science has always been, in Newton's phrase, 'mathematical and experimental philosophy': many scientists, and some positivist philosophers would want to deny the name of science to any part but this. But there has in practice always been the other strand to be discerned, drawing a certain prestige lately from association with its highly successful partner—this is the strand traditionally referred to as 'natural theology'. All I have been doing in this essay is applying that distinction to certain things in the science, not of the past, but of our own day.

The Creation, the Apocalypse, the Foundations of Morality, the Justification of Virtue: these are problems of perennial interest, and our contemporary scientific myths are only one more instalment in

2. H. Butterfield, in the *Cambridge Journal*, Vol. I, p. 8 (1948).

the series of attempted solutions. So next time we go into an eighteenth-century library, and notice these rows of sermons and doctrinal treatises lining the shelves, we need not be puzzled by them. Now we are in a position to recognize them for what they are: the forerunners, in more ways than one might at first suspect, of the popular science books which have displaced them.

PART TWO

A CONSIDERATION
OF COSMOLOGISTS

Arthur Koestler I

This is a bad time for polymaths. The old jibe about a Jack-of-all-trades being master of none has bitten deep into our minds, so that few people will admit to an intellectual grasp of anything more than a narrow range of experience. In a fragmented culture, everybody is *expected* to be a specialist: so men cling to the professional standards of their guilds as the lifebelts which will keep them afloat on a sea of general ideas which they have lost the capacity either to swim in, or to plumb. By now, a politician who habitually quoted Horace in Parliament would be endangering his position as an M.P.—and what matters more, it would be the same if he had a reputation for habitually quoting (say) T. S. Eliot. Natural scientists, again, look with suspicion at those colleagues who stray too far outside their specialisms, and for the most part they shut their eyes to the very existence of philosophy. In return, the philosophers have made their craft a "profession" of its own: a professional philosopher no longer needs to understand even the broadest ideas of contemporary natural science—to say nothing of its factual discoveries.

To use a phrase of Pascal's, ours is an age dominated by the *esprit géometrique*. Narrow precision and deductive exactitude carry the palms: analogies are distrusted, virtuosity suspect. So to embark on any large synthesis of the different sciences—still more, to range with confidence through both the sciences and the humanities—a man must have both a level head and a well-stocked mind. His *esprit géometrique* must be counterbalanced by a well-developed *esprit de finesse*, and his personal position must be so

Originally published as "Koestler's Act of Creation" in *Encounter* (July 1964).

well assured that he need no longer be afraid of making mistakes.

Who among our contemporaries dares measure himself up against this specification? One man, at any rate, has had the courage—or the foolhardiness—to do so: Mr. Arthur Koestler, whose remarkable new book *The Act of Creation* is an attempt to integrate into a single system of ideas the results of modern physiology, psychological theory and his own novel analysis of artistic creativity and scientific discovery.

How is one to appraise a book of this kind? Some readers will be tempted to burke the issue by retorting, "Who is Arthur Koestler to write such a book, anyway?" (This will be especially tempting for the theoretical psychologists—such as Skinner of Harvard—on whom Koestler is most severe.) Yet this would be scarcely fair; for who, we might reply in return is *better* qualified than Koestler to cover the whole of his chosen field? To the general reader no doubt, *Darkness at Noon* is "typical". Koestler, and *The Sleepwalkers* was a curious sideline—the propagandist storyteller being distracted by curiosity into the world of scholarship. Yet, although Koestler won his major reputation as a political novelist and has since become widely known as a public figure, his initial training was in fact scientific, and he started his career as science correspondent for the Ullstein newspapers in Berlin.

Indeed, one might try looking at Koestler's overall achievement in the opposite perspective; for perhaps it will appear in retrospect that the curious, anomalous phase in his career was that when he was writing novels, instead of exercising his true concerns as a "natural philosopher." Certainly, many of those who know him personally will feel that *The Act of Creation* gives a truer picture of his actual cast of mind than the nightmare-obsessed plots of his earlier stories. At last the shadow of the Stalin-Hitler Era can be forgotten, the menace of the Cold War is lifting, and Koestler can turn his ample, serious mind away from the tortuousness of political conspiracy, to view with all-embracing scope the larger sphere of Man and Nature. Scientific and philosophical reflections which he sketched fifteen years ago in the book *Insight and Outlook*, and which he focused through the lens of Johann Kepler's personality in *The Sleepwalkers*, are expounded at length for the first time in the 750 pages of his new book. Throughout the whole close-textured volume, there is not a single reference to communism, Karl Marx is alluded to only once, *à propos* of Darwin, and there is one passing reference alone to the Spanish Civil War; for here, without the slightest ambiguity, Arthur Koestler is staking his claim to be regarded *as a scientist*. Whatever one's final verdict on his conclusions, it is at any rate essential to look at his argument carefully and seriously.

The Act of Creation is scarcely the kind of book one can summarise. At best a critic can describe and paraphrase its argument in general terms, before attempting to assess its points of strength and weakness. To begin with, then, Koestler's exposition is divided into two halves, which could—echoing Parmenides—be labelled "the way down" and "the way up." The first half of the book ("The Art of Discovery and the Discoveries of Art") presents a general theory about the nature of creative originality as displayed in humour, natural science, and the fine arts. The second half ("Habit and Originality") argues that this creativity has analogies on all levels of an organic hierarchy which extends from the sub-cellular world of molecular genetics, up through the spheres of embryology and physiology to those of human behaviour, learning, and symbolic expression. The first half of the book is not (as one might guess) a popular presentation of the same theories which the second half expounds in more technical language: the two parts are strictly complementary, and both are equally necessary for an understanding of Koestler's whole thesis.

This thesis is a threefold one. Koestler hopes to demonstrate, in the first place, that "organic life, in all its manifestations, from morphogenesis to symbolic thought, is governed by 'rules of the game' which lend it coherence, order, and unity-in-variety"; secondly, that "these rules . . . whether innate or acquired, are represented in coded form on various levels, from the chromosomes to the structures in the nervous system responsible for symbolic thought"; and finally, that all creative originality is the product of what he calls "bisociation," which involves "the combination, re-shuffling and re-structuring of skills" or other rule-governed activities. Artistic and intellectual novelties all spring from the fusion—indeed from the *marriage*—of separate pre-existing routines: in a manner of speaking, then, "the basic model of the creative act" is "the bisociation of two genetic codes" in sexual reproduction, since this too is the ultimate source of novelty on the organic level.

Let us begin by looking at Koestler's initial claim, to identify a common pattern of originative activity shared by the jester, the sage (or scientist), and the artist; then go on to consider the "organic hierarchy" which he sees as underlying all kinds of creative activity; and finally ask whether he is justified in equating the end-points of his two investigations—human creativity, on the one hand, and top-level organic activity, on the other.

For much of the first half of the book, Koestler's *esprit de finesse* is at its best. What he does here is to set out for us a *theory* about the major originative activities of the human mind; and he shows us the force of this new view by a subtle presentation of selected examples which gradually insinuate his key-idea into our minds. This idea is a *theoretical* one, in a quite strict sense of that term. When giving a completely general account of *motion*, for example, physicists start by presenting a simplified "ideal type" of the phenomenon in question, from which all the complicating factors (*i.e.*, "forces") have been eliminated by a deliberate abstraction. Such "inertial motion" need never in fact take place. We are presented with the model solely as an object of intellectual comparison, and all the motions that we actually observe are explained as deviations from this ideal, brought about by different permutations and combinations of physical forces.

Koestler's procedure is similar. He, too, treats scientific and artistic originality as difficult and complex phenomena, which are best explained, not by looking at them directly, but rather by comparing them with a simpler, idealised example from which complexities have been eliminated. This simplified example he finds in the activity of the *jester*: as he explains on making the transition "from humour to discovery. . . ."

> I have started this inquiry with an analysis of humour because it is the only domain of creative activity where a complex pattern of intellectual stimulation elicits a sharply defined response in the nature of a physiological reflex.

What the scientist and the artist do to us in subtle, delicate and varied ways, the jester does to us abruptly; yet for all the apparent simplicity of his achievement it already carries within itself the leading characteristics of its more adult counterparts—

> *Originality* or unexpectedness; *emphasis* through selection, exaggeration and simplification; and economy or *implicitness* which calls for extrapolation, interpolation and transposition.

Above all, the originality of the humorist displays in simplified form a pattern which Koestler is going to trace out for us in all other spheres of human and organic activity:

> The pattern underlying all varieties of humour is "bisociative"—perceiving a situation or event in two habitually incompatible associative contexts. This causes an abrupt transfer of the train of thought from one matrix to another governed by a different logic or "rule of the game." But certain emotions, owing to their greater inertia and persistence, cannot follow such nimble jumps of thought; discarded by reason, they are worked off along channels of least resistance in laughter.[1]

Notoriously, theories of humour tend to be dreary reading. Fortunately, Koestler has a much better sense of humour than (say) Freud, whose account in some respects anticipated his own, so we are spared almost entirely the embarrassing feeling that somehow or other—in the course of being "explained"—the whole point of the jokes under discussion has quietly evaporated. If this is so, it is because Koestler does more than merely identify the *locus* of a joke as the intersection of two "planes." He also analyses with great sensitivity the emotional mechanisms by which this kind of juxta-

---

1. What are these "matrices," which the jester "bisociates"? Here is Koestler's general account of them:

"The term 'matrix' was introduced to refer to any skill or ability, to any pattern of activity governed by a set of rules—its 'code.' All ordered behaviour, from embryonic development to verbal thinking, is controlled by 'rules of the game,' which lend it coherence and stability, but leave it sufficient degrees of freedom for flexible strategies adapted to environmental conditions. The ambiguity of the term 'code' ('code of laws'— 'coded message') is deliberate, and reflects a characteristic property of the nervous system: to control all bodily activities by means of coded signals. . . ."

position produces its effect on us, relating them back by stages to the laughter of the infant in the cradle, who responds to a mock attack with the same laughter that actual tickling would provoke.

With this preamble behind us, we read on eagerly, to see how Koestler will apply the concept of "bisociation" in the more contentious areas of science and the arts. The closing words of the initial section provide a text for what follows—

> Habits are the indispensable core of stability and ordered behaviour; they also have a tendency to become mechanised and to reduce man to the status of a conditioned automaton. The creative act, by connecting previously unrelated dimensions of experience, enables him to attain to a higher level of mental evolution. It is an act of liberation—the defeat of habit by originality.

High-grade intellectual and artistic activities, he argues, display the same contrast between "automatised routines" and moments of "creative originality," prompted by the sudden working-together of "previously unrelated dimensions of experience"; and, in his account of the evolution of scientific ideas, he applies this fundamental contrast convincingly and illuminatingly. For, indeed, the development of science has been characterised by just such an alternation between periods of calm, straightforward growth—during which the adequacy of the fundamental concepts and techniques was not seriously in question—separated by briefer phases of intellectual metamorphosis, during which those fundamental concepts or techniques had to be re-thought entirely. And each such transformation has been preceded by a period of frustration, during which scientists were psychologically "blocked," and hunted around in vain for a way of escape from apparently insoluble difficulties.

In this section, one is especially grateful for the delicacy and clarity with which Koestler describes the actual process of intellectual discovery—in a word, for his *finesse*: this is shown, for instance, in his thumbnail sketches of Gutenberg's invention of movable type, Kepler's speculations about gravity, and Darwin's discovery of "natural selection." By reconstructing the intellectual situations confronting these great innovators, he is able in the end to lend great plausibility to a claim that one might otherwise have resisted—namely, that there is, psychologically speaking, an unbroken spectrum which stretches from the sublimity of a Kepler or a Newton mastering the movement of the planets, right the way down to Köhler's chimpanzees laying siege to a banana. At both ends of the scale, as Koestler demonstrates, "discovery" comes about when techniques earlier developed quite separately are, for the first time, combined together to surmount a new obstacle; and the same patterns of frustration (or "blocking"), withdrawal into the imagination, hunting around and sudden insight can be observed equally in the ape and in the mathematician.

If science strikes the contemporary literary mind as essentially boring, that is because the crucial importance of imagination and insight in scientific discovery has too often been played down. Somehow, the routine phases of scientific consolidation have come to be regarded as more respectable and "rational." Yet, as Koestler insists,

> the aesthetic satisfaction derived from an elegant mathematical demonstration, a cosmological theory, a map of the human brain, or an ingenious chess problem, may equal that of any artistic experience—given a certain connoisseurship. . . .
>
> To derive pleasure from the art of discovery, as from the other arts, the consumer—in this case the student—must be made to re-live, to some extent, the creative process. In other words, he must be induced, with proper aid and guidance, to make some of the fundamental discoveries of science by himself, to experience in his own mind some of those flashes of insight which have lightened its path. This means that the history of science ought to be made an essential part of the curriculum, that science should be represented in its evolutionary context—and not as a Minerva born fully armed. It further means that the paradoxes, the "blocked matrices" which confronted Archimedes, Copernicus, Galileo, Newton, Harvey, Darwin, Einstein should be reconstructed in their historical setting and presented in the form of riddles—with appropriate hints—to eager young minds. . . .
>
> The creative achievements of the scientist lack the "audience appeal" of the artist's for several reasons briefly mentioned—technical jargon, antiquated teaching methods,

cultural prejudice. The boredom created by these factors has accentuated the artificial frontiers between continuous domains of creativity.

Koestler's own picture of scientific discovery is not just more engaging or "human" than usual—for we all know those terrible coy anecdotes thrown in, like currants, to give flavour to stodgy books about "the great scientists"—but, rather, more gripping and compelling, proving in action his claim that we can all receive the intellectual quandaries of our forefathers. Yet in one respect his theory is, in my opinion, open to serious question. I am not referring to his errors of historical fact and interpretation: some of these are inevitable in a book of such length, and they scarcely touch Koestler's central thesis.[2] What I do query is the importance he attaches to the moments of insight being brief and sudden: he speaks repeatedly of "*flashes* of insight," "the bisociative *click*," and so on. Now, there is no doubt that discoveries have sometimes happened as a result of sudden—even instantaneous—inspirations, so that Henri Poincaré (for instance) could actually time the moment at which he realised that the Fuchsian functions that he had been studying mathematically were identical with those of non-

---

2. For the sake of improving the inevitable second edition, let me just note the following: (1) the author of *L'Homme Machine* (p. 48) was Julien de la Mettrie (1709−51); (2) the second law of thermodynamics antedated Maxwell (p. 128); (3) Anaximander and Lucretius did *not* believe in the evolutionary development of one species into another (pp. 131, 137); (4) Darwin had almost certainly hit on the idea of "natural selection" *before* reading Malthus (p. 140); (5) it was not generally believed by "the educated classes" in the 15th century A.D. that the earth "was a flat disc, of a rectangle perhaps" (p. 227, but cf. p. 255); (6) Newton was explicitly aware that light has wave-like characteristics as well as corpuscular ones (p. 240); (7) the status of Wöhler's synthesis of urea is nowadays in serious doubt (p. 240); and (8) the man who jumped into the crater of Etna to gain immortality was not Eudoxus of Cnidus but Empedocles of Akragas (p. 256).

Note also: the same quotation is attributed to Galton on p. 160 and to Taine on p. 165; T. H. Huxley is credited with a remark of Herbert Spencer's on p. 214; George Sarton is called Henry on p. 224; Miletus is called "Milos" on p. 227; and Hipparchus is misspelt "Hypparchus" on p. 234. Likewise on p. 588, for "Ecclesiastes" read "Epicurus."

I shall return later to the question why Koestler is unable to extend to Aristotle the intellectual charity he insists on in the case of all other scientists—*cf.* the remarks about Aristotle's "absurd theory of physics" which "paid no attention to quantity or measurement" and was "full of glaring self-contradictions."

Euclidean geometry to the instant at which he put his foot on the step of an omnibus at Coutances. Equally, some scientific discoveries have been born from reverie as surely as Coleridge's *Kubla Khan*—e.g., Kekulé's theory of molecular rings. But surely to treat this kind of discovery as the typical case, rather than an exceptional one, is to make one's account of scientific discovery over-histrionic. It is, no doubt, a matter of great significance in that the central novelty in a new theory *can* suggest itself to the scientist as a result of a momentary association of ideas; but more generally discovery is what Koestler would call a "diluted Eureka process," and for him to insist on the suddenness of scientific inspiration makes one fear that for once he is allowing the novelist in him to get the better of the psychologist. After all, as he concedes, flashes of inspiration may be deceptive, and the all-embracing term "discovery" can scarcely be used with accuracy to refer to anything shorter than the whole establishment of a new concept. Ideas may come to us in a flash: discoveries must be earned laboriously.

When this amendment has been made, we can discount very largely the element of literary exaggeration to be found in such a passage as this:

> Yet the evidence for large chunks of irrationality embedded in the creative process, not only in art (where we are ready to accept it) but in the exact sciences as well, cannot be disputed; and it is particularly conspicuous in the most rational of all sciences: mathematics and mathematical physics . . . a branch of knowledge which operates predominantly with abstract symbols, whose entire rationale and credo are objectivity, verifiability, logicality, turns out to be dependent on mental processes which are subjective, irrational, and verifiable only after the event.

Over the question where we get new ideas from, questions about "logicality" and the rest scarcely arise. We may come by them suddenly, we may work our way to them painfully: it is the total process of discovery—verification after the event and all—which has to be judged by the canons of rationality. Koestler's false antithesis between irrational, inspirational discovery and rational, routine consolidation—God-given originality and earthbound habit—is thus a romantic exaggeration, rather than a piece of serious psychology.

So we move on to the fine arts. Here again, Koestler's sensitivity and range of experience are most rewarding. In many cases, he is able to demonstrate in the field of the arts, also, patterns of creation and originality analogous to those he expounded in his analyses of humour and scientific discovery. The resemblances remain striking, for just so long as he takes care to talk about genuine "originality" in the fine arts—as contrasted with "creativity," in a more general sense. He demonstrates very justly that the same kind of "evolution of ideas" that can be traced out in the history of science can be recognised equally in the history of arts.

> In both fields the truly original geniuses are rare compared with the enormous number of talented practitioners; the former acting as spearheads, opening up new territories, which the latter will then diligently cultivate. In both fields there are periods of crisis, of "creative anarchy," leading to a break-through to new frontiers—followed by decades, or centuries of consolidation, orthodoxy, stagnation, and decadence—until a new crisis arises, of holy discontent, which starts the cycle again.

Just so long as he concerns himself with stylistic innovations, one can see how Koestler's concept of "bisociation" helps to explain the nature of creative originality, in art as much as in science. Equally, the process of frustration, withdrawal, imagination, and insight leading to a new step forward, which was evident in many of his scientific examples, can once more be shown taking place in the minds of those creative artists who have broken with earlier styles and media, and branched out into new and original directions. (Is not to-day's great novelty in painting—Pop Art—well described as a "bisociation" of the pre-existing "matrices" of abstract painting and the mass media?)

Yet in this section, too, Koestler the romantic gets the better of Koestler the psychologist and, as we read, the idea of bisociation gradually loses its initial precision. To begin with, the word had been introduced as a straightforward single-purpose term, whose relevance to any particular example was clear enough. The wine-press and the seal are bisociated to yield moulded lead type (Gutenberg); abstract collage and advertising material are bisociated to yield Pop Art. But Koestler is not content with a good, but restricted theory of artistic *innovation*; what he wants to do is to give an account of all artistic *creation* conforming to the same pattern. Yet how is this to be done? When we turn to consider a run-of-the-mill string quartet or poem, painting or novel, involving no particular stylistic innovations, what are we in that case to say is "bisociated" with what?

For the next fifty pages, Koestler keeps the equation "Creativity = Bisociation" on its feet only by desperate measures, and there are moments when he veers perilously near to literary humbug. In a play, for instance, the bisociation is (he claims) between the actor's real personality and his stage-part, in a painting between "sensory qualities and emotive potential," in a great novel between "incompatible frames of experience or scales of value, illuminated in consciousness by the bisociative act" or more portentously between "the Trivial Plane or everyday life" and "the Tragic Plane . . . the essential solitude of man." In due course, he even uses the term "bisociation" to express his own species of philosophical idealism—

> Man always looks at nature through coloured glasses—through mythological, anthropomorphic, or conceptual matrices—even when he is not conscious of it and believes that he is engaged in "pure vision," unsullied by any meaning.

Thus "form" and "meaning" too are made to serve as a "bisociated" pair, and the way is paved for a general theory of artistic expression owing something to Jung, and something to Bergson and the idealists:

> The difficulty of analysing the aesthetic experience is not due to its irreducible quality, but to the wealth, the unconscious and non-verbal character of the matrices which interlace in it, along ascending gradients in various dimensions. Whether the gradient is as steep and dramatic as in a Grunewald or El Greco, or gently ascending through green pastures, it always points towards a peak—not of technical perfection, but of some archetypal form of experience. . . .
>
> The aesthetic experience aroused by a work of art is derived from a series of bisociative processes which happen virtually at once and cannot be rendered in verbal language without suffering impoverishment and distortion.

This position in philosophical aesthetics combines with his earlier psychological theory of original discovery to yield the following general thesis about the creative capacities of the human mind:

> The *locus in quo* of human creativity is always on the line of intersection between two planes; and in the highest forms of creativity between the Tragic or Absolute, and the Trivial Plane. The scientist discovers the working of eternal laws in the ephemeral grain of sand, or in the contractions of a dead frog's leg hanging on a washing-line. The artist carves out the image of the god which he saw hidden in a piece of wood. The comedian discovers that he has known the god from a plum tree.
>
> This interlacing of the two planes is found in all great works of art, and at the origin of all great discoveries of science. The artist and scientist are condemned—or privileged—to walk on the line of intersection as on a tightrope.

This message is, no doubt, very edifying, but by this time we have left Koestler's psychological theory about the nature of originality far behind. In the process, the very useful term "bisociation" has become fuzzed and blurred, until it has lost most of its explanatory merit. To me, at any rate, this is a sad disappointment. At the outset, Koestler had seemed to be proclaiming for the psychology of invention what Lavoisier declared about chemistry—that "it is time to bring it to a more rigorous way of reasoning," and to abandon the loose and undisciplined linguistic habits characteristic of earlier speculation. In the course of his section on art, however, Koestler's love of generalisation leads him to devalue his own new coinage, to a point at which I find myself echoing Lavoisier's protest about the chemist's use of the term "phlogiston"—

> A vague principle, which they in no way define rigorously, and which in consequence is adaptable to any explanation they please. Now this principle has weight, and again it is weightless; now it is free fire, and again it is fire combined with the element earth; now it penetrates right through the pores of vessels, and again it finds bodies impenetrable. It

explains at once causticity and its opposite, translucence and opacity, colours and the absence of colours. It is a veritable Proteus, changing form at every instant.

Then suddenly, at the end, Koestler brings himself back to artistic originality proper, as displayed in the history of art, and we find ourselves back on beam again. The pity is that he does not take sufficient care to distinguish, throughout, between the creative capacities which any artist must have, in the nature of his profession, and the special originativeness which differentiates those men who create new styles or exploit new media. Even a great artist is necessarily for the most part a craftsman; and it is, once again, a romantic fancy to insist on finding "bisociative processes"—in the sense in which Koestler originally defined that phrase—in all artistic creation whatever.

Here we must leave Koestler's account of human creativity at the highest level, and turn to the general system of biological and psychological concepts in which he seeks to root that account. This occupies the second half of *The Act of Creation*. Once again, it is scarcely possible to summarise his argument, especially as in this second half much of the discussion is at rather a technical level. Still, one can perhaps once more describe and paraphrase the case he puts forward—even though one can do no more than hint at the richness of examples and illustrations which form so impressive a part of his exposition.

The starting-point in this second half is the same as in the first: the idea of a "matrix" characterised by certain "rules of the game." Matrices of this kind (Koestler claims) are to be found in all the manifestations of organic life "from embryonic development to symbolic thought"—

These rules or codes, whether phylogenetically or ontogenetically acquired, function on all levels of the [organic] hierarchy, from the chromosomes to the neuron-circuits responsible for verbal thinking. Each code represents the fixed, invariant aspect of an adaptable skill or matrix of behaviour. I shall take the stylistic licence of using the word "skill" in a broad sense, as a synonym for "matrix," and

> shall speak of the morphogenetic skills which enable the egg
> to grow into a hen, of the vegetative skills of maintaining
> homeostasis, of perceptual, locomotive, and verbal skills.

(We shall have to ask later whether this very broad use of the word
"skill" involves only a *stylistic* licence.)

Starting from this point, Koestler embarks on a thumbnail
restatement of the leading discoveries of 20th-century biology and
psychology. At every stage he underlines the contrast between
fixed, predetermined patterns of growth and activity, and that
residual scope for individual variability and interaction with the
environment that biologists call "plasticity." In a newly-con-
ceived, single-celled embryo, the "genetic code" transmitted by
the DNA molecules which make up the nucleus of the cell serves in
many respects to govern and limit the possible ways in which the
resulting organism can grow, develop, and behave. To that extent,
so to speak, the embryo already has certain in-built physiological
and psychological "habits," and its scope for "originality" is
thereby curtailed. (The qualification "so to speak" is important.)
Yet the existence of these fixed patterns of activity on a lower
organic level is in practice a necessary condition, if there is to be
any real chance of freedom or development at a higher physiological or psychological level. As Koestler puts it,

> the emergence of life means the emergence of spontaneous,
> organised exertion to maintain and produce originally un-
> stable forms of equilibrium in a statistically improbable sys-
> tem in the teeth of an environment governed by the laws of
> probability.

(An example may help: at the physiological level, a spastic
cannot achieve anything like the same degree of muscular control
as a normal human being, and his behaviour is therefore subject to a
much greater degree of physiological variability; yet on that
account he must be judged—at the behavioural level—to have *less*
scope for freedom and originality, rather than *more*.)

Koestler pays particularly close attention to the facts of embryological development and physiological regeneration. As he empha-
sises, one of the fundamental capacities of the organism is the
ability when necessary to revert to an earlier stage of development
(*reculer pour mieux sauter*), so as to re-create or make good the

loss of some injured limb or organ. In this, running ahead somewhat, he sees an image pre-figuring a spiritual phenomenon—

> the perennial myth of the prophet's and hero's temporary isolation and retreat from human society—followed by his triumphant return endowed with new powers. Buddha and Mohammed go out into the desert; Joseph is thrown into the well; Jesus is resurrected from the tomb. Jung's "death-and-rebirth" motif, Toynbee's "withdrawal and return" reflect the same archetypal motif. It seems that *reculer pour mieux sauter* is a principle of universal validity in the evolution of species, cultures, and individuals, guiding their progression by feedback from the past.

As to this biological phase in Koestler's argument, let me add only one further qualification: though the analogies and resemblances for which he is arguing are in their way quite appealing, he is undiscriminating in the evidence which he quotes to establish them. Though he himself protests later on (p. 556) against applying to human beings psychological conclusions derived from the study of rats and pigeons, for instance, his embryological argument jumps without comment between observations on salamanders, frogs, and human embryos (*cf.* pp. 427−8). We are not, in fact, in a position to *know* at the present moment whether this is entirely legitimate or not.[3] But I do not press the point, for this biological section, like the corresponding section on "The Jester" in the first

3. Incidentally, embryology is not a field in which one can safely mix present-day evidence with quotations derived from sources thirty years old or more. In particular, the term "organiser," which Koestler uses without comment, is by now under a cloud among professional biologists—and for good reason—and it is not safe to use the word as though it were strictly equivalent to "evocator" or "inducer." So Koestler's text at this point needs to be read with care. A reference taken at second hand from a book of Michael Polanyi's to a paper published by C. H. Waddington in 1932 hardly rates as highly as an encyclopædia article by Jean Brachet dated 1955, and one would have preferred to see the argument based on—say—Waddington's own post-1960 work. Incidentally, some straightforward corrections are again required: *e.g.*, the formation of two complete frogs from the separated halves of a frog-egg and the production of rudimentary kidneys from individual cells taken from the kidney area are *not* examples of the same phenomenon as the re-formation of a living sponge from its dissociated cells (pp. 423−4).

part, is intended only as a preamble to the more serious discussion of psychology and behaviour.

Let me say at once that much of Koestler's long and careful discussion of contemporary psychological theory is valuable and pertinent. For the last half-century theoretical psychologists have been as deeply divided into rival schools and "tendencies" as metaphysicians or theologians have ever been. Of the whole field which Arthur Koestler has set himself to explore in this book, no single part has been more in need of attention from a sympathetic, unprejudiced, and eclectic observer. In a warmly-phrased foreword Professor Sir Cyril Burt commends Koestler's attempt to bring together and integrate into a single consistent scheme the separate results which have been achieved by psychologists of all these different tendencies. And in this respect his praise—so far as I am in a position to judge—appears well deserved. At this stage Koestler turns away from his attempt to generalise the notion of "bisociation." Instead, he makes it his business to do something rather different—namely, to show how all the great and apparently disorderly variety of human behaviour and activities can be ordered as a hierarchy of skills, in which those at the higher level presuppose, and are dependent upon, the previous acquisition of those at the lower level. From the moment of birth, when our instinctual skills and capacities are—necessarily—our only psychological equipment, we gradually acquire more and more complex abilities: the ability to "perceive" (for it is important to realise that perceiving *is* something we "do"), the ability to generalise and discriminate, the ability to recognise causes, etc. Furthermore, Koestler shows, the capacity to move on from any one level to the next is bound up with the possession of an "inherent exploratory drive," and apart from this the behaviour even of rats and pigeons (to say nothing of human behaviour) cannot be properly understood. This is the point at which Koestler makes his most important and original contributions to psychological theory, for exploratory and creative activities are precisely those with which recent psychological theories have been least successful in dealing.

> It took experimental psychology nearly fifty years to re-discover, after its Dark Ages of need-reducing S.R. [stimulus-response] theories, that rats and men are pleasure-seeking creatures, that some activities are pleasurably self-reward-

ing, and that exploring the environment, solving a chess problem, or learning to play the guitar are among these activities.

Those who could envisage only a pattern of external stimuli producing behavioural responses were at a loss to explain (say) how a rat sometimes learns to run through a maze without actually having this learning "reinforced" by punishments or rewards—

> Several writers have considered the possibility that [this kind of learning] comes from the reduction of curiosity. . . . One might as well say that composing a song is a silence-reducing activity.

All this has a considerable bearing on our understanding of human behaviour, and leads to a system of ideas more comprehensive and adequate than we get from either pure learning-theory or pure *Gestalt* theory:

> On the elementary levels of learning a skill a varying amount of stamping-in is required, depending on the organism's "ripeness" for the task; or, to put it the other way round, depending on the "naturalness" of the task relative to the organism's existing skills. Learning to type requires more stamping-in than learning to ride a bicycle; the former is comparable to the blindfold memorising of a maze, the latter to the gradual adjustment of the various interlocking servo-mechanisms. . . . But even in acquiring a mechanical skill like typing, bit-by-bit learning plays in fact a lesser part than seems to be the case. The typist's mental map of the keyboard is not simply a rote-learned aggregation of twenty-six letters (plus numbers and signs) distributed at random; it is a "coded" map, structured by a system of co-ordinates—the resting position of the fingers—and by the frequency-rating of letters, syllables, etc. . . . Whole-learning invades bit-learning at every opportunity; if the meaningless is to be retained, the mind must smuggle meaning into it.

At every level, accordingly, we build a kind of "floor" of routine activities, which gives meaning to one whole range of concepts; then, in turn, this "floor" serves as the foundation on to which a higher range or "storey" of activities is built, with its own more sophisticated range of concepts and meanings. And the step by which we pass up from one floor to the next above is neither a pure process of trial-and-error, nor a brand-new intellectual leap— something coming entirely "out of the blue," without any previous preparation. It is, Koestler argues, a genuine act of *discovery*, in which we "put two and two together"—or, as Hebb puts it, "our evidence thus points to the conclusion that a new insight consists of a *re-combination of pre-existent mediating processes*, not the sudden appearance of a wholly new process."

Finally, we are back again at the level of thought, language, and symbolism in general. Koestler's last five chapters present a theory of thinking intended to link together the account of originality contained in the first half with the ideas about biology and psychology making up most of the second. Here, I confess—and this may be my fault as much as his—I find myself once again losing sympathy with him. A general framework of psychological theory such as he had outlined during the previous hundred pages could naturally have been extended into the linguistic field in such a way as to produce something valuable and important for philosophy as well as for psychology. Indeed, the pattern of analysis he adopts for characterising pre-linguistic and non-linguistic behaviour has genuine and significant analogies to the pattern of analysis used by Ludwig Wittgenstein in his later years to describe the linguistic and symbolic activities which give "meaning" to our concepts. The "rules of the game" associated with the "coding" of Koestler's behavioural "matrices" immediately call to mind the rules of the "language-games" in terms of which Wittgenstein explained how our linguistic symbols are related to nonlinguistic "forms of life"; and the same sort of hierarchy that Koestler recognises in prelinguistic behaviour was recognised and described by Wittgenstein as an essential feature (or structure) within the pattern of linguistic activities. To give one specific example: Koestler's whole discussion of Köhler's experiments on the counting abilities of lower animals and birds and the relation of these experiments to the human capacity to calculate symbolically, gains a new depth and interest if read in conjunction with Wittgenstein's discussion of the same subject in *The Brown Book*.

I say this, not out of professional concern as an "analytical philosopher," nor out of personal loyalty to Wittgenstein, but rather because, through neglecting the results achieved by recent philosophical analysis in Britain and America, Koestler is led to take up some unnecessarily naïve philosophical positions. His insistence that the behaviour-patterns associated with our linguistic concepts are fully developed before the relevant *word* becomes attached to them is very just. But in expounding the implications of this important point he places too much weight on an over-simple "reference" theory of meaning—words for him are "labels," which acquire meaning through being used to "refer to" things. This section of his book would have been more illuminating if he had taken more notice of the central insight of contemporary analytical philosophy: namely, that "labelling" and "reference" represent only one of the many different ways in which words become associated with behaviour-patterns in the course of our conceptual and mental life. (There is no obvious reason why he should have rejected this particular refinement: from time to time, indeed, his actual style and choice of examples recalls Gilbert Ryle.) It seems to be his obsession with neurology—which makes him speak of "the neuron-circuits *responsible for* verbal thinking" where others might prefer to say "correlated," or "associated with"—that leads him to prefer a crude and muddled account of concept-formation written by a neurosurgeon who is, philosophically speaking, little further on than Locke or Berkeley to the more sophisticated and discriminating accounts of the same phenomenon that he could have got from contemporary analytical philosophers.

Can one pass any general verdict on *The Act of Creation* as a whole? It is too soon for a final judgment, but let me do my best. As in the first half, the second half of the book contains a substantial amount of novel and illuminating material flawed by the same aspiration to be all-embracing. In the first part, Koestler started with an admirable theory of originality, but forced it beyond its proper scope so as to make it serve as a completely general account of creativity. In the second part he began with a clear idea about how to weave together harmoniously the ravelled strands of theoretical psychology, but once again he was tempted to extend the resulting notion of a "hierarchy" too far—aiming at a single system reaching downwards to embrace embryology and bio-chemistry, and upwards to a union with scientific and artistic originality. In both cases, his fundamental concepts will not stand up to the strain he puts upon them. Just as his notion of "bisociation" ended up by being as vague as "phlogiston," so now his "organic hierarchy of activities," each with its own "codes" and "rules of the game," ends up as a mere aggregate of partial analogies. It is always the death of a theory to be made to *do too much*—to be stretched and stretched until it can explain absolutely anything. When criticising other people's arguments Koestler sees this point clearly enough: when he remarks of the term "regeneration" that it is "difficult to find a satisfactory definition which would embrace the whole range of phenomena to which the term is applied," or dismisses Pavlov in the words, " 'Conditioning' is still a useful term when applied to induced changes in glandular and visceral reactions, but leads to confusion when used in a loose, analogical way for other types of learning." It is not easy to exonerate Koestler's use of the terms "bisociation," "matrix," "code," and "rule" from the same charge.

The trouble is that, once the necessary steps are taken to escape this particular charge, *The Act of Creation* simply falls apart. It was written as two separate books, and it is in fact two separate books. As things stand, the two halves do not have much to do with one another, though—to do Koestler justice—it is possible that the connections he would like to see between them may be better established a hundred years hence. At the present time, at any rate, the central conceptual spine around which Koestler has organised his argument embodies not so much a scientific theory as a philosophical vision of nature. It may eventually prove legitimate to equate the mechanisms of physiological and biochemical genesis, the development of new routines of behaviour at all levels, and the creative originality of human artists and scientists—and to embrace them all under a single concept. But, with all respect, I must say that Koestler has not convinced me that this time has yet come. For the moment, the things in his new book to which I shall turn again with pleasure and fascination are, first, his sensitive and humane reconstructions of the intellectual quandaries facing earlier scientists (except the unsympathetic Aristotle) and, second, his re-ordering of psychological theory into a hierarchy of habits and novel discoveries.

What Koestler gives us here is a vision of nature. One can go further, and add that it is an essentially *romantic* vision of nature. Koestler set out to stake his claims as a scientist. But he proves to be, rather, a natural philosopher in the manner of Goethe. In several ways, indeed, the system of ideas which he has been developing in his own mind from the days of *Insight and Outlook*, through *The Sleepwalkers*, to *The Act of Creation* resembles Goethe's own natural philosophy. This fact shows itself not merely in Koestler's positive theses, but equally in his prejudices and hostilities. For, as so often, the overall direction of his thought shows itself particularly clearly in his polemical passages, where he takes a dogmatic stand in order to preserve his system of ideas more rigidly than our present understanding really justifies.

For example, Koestler shares with earlier romantics and idealists an exaggerated suspicion of formal logic. ("The so-called law of contradiction . . . is a late acquisition in the growth of individuals and cultures. . . . The unconscious mind, the mind of the child and the primitive, are indifferent to it. So are the Eastern philosophies which teach the unity of opposites, as well as Western theologians and quantum physicists.") The intrusion of mathematics into science, and the reliance on intellectual abstraction have also, in his view, been overdone. In the circumstances it is perhaps surprising that he is quite so venomous about Aristotle, with whose dynamic view of nature Goethe and the subsequent Romantics had a good deal in common. Perhaps it is only Aristotle's *Physics* that Koestler rejects—though I, for one, would be happier if he had produced for inspection just *one* of the "glaringly evident self-contradictions" he finds in it. (Be as charitable in interpreting Aristotle as Koestler is towards Kepler, and these deficiencies will soon disappear.)

It is understandable, of course, that Koestler should be merciless in rejecting psychological behaviourism: this is one of the main targets against which he produces some very solid and satisfactory arguments. His attacks on neo-Darwinism, on the other hand, are more modish and cliquey. Like Michael Polanyi, Marjorie Grene, and Gertrude Himmelfarb, he begins by caricaturing contemporary evolutionary genetics, and then pours scorn on his own caricature. (I know of *no* evolutionist who teaches that "the courtship and fighting rituals of various species have all evolved by 'pure chance,' " and *all* evolutionists recognise that evolution by natural selection frequently involves *pre*-adaptation to a changing environment.) What I do think Koestler needs to do is to reconcile his *rejection* of chance, as a principal factor in organic evolution, with his positive *espousal* of it as a factor in discovery (*cf.*, "the familiar pattern of a playful habit being connected with a blocked matrix, with chance acting as a trigger").

In one respect above all, Koestler's attempt to wed 20th-century science and philosophical romanticism into a unified natural philosophy ends by producing intellectual difficulties. These show themselves in his highly-ambivalent attitude towards "mechanistic" ideas. As a romantic, he must surely be as opposed to a purely mechanical view of human nature as he is to evolution-through-pure-chance and to mathematical reductionism. Sure enough, he rejects Francis Galton's teaching that the chains of association by which ideas are brought into our minds act "in a mechanically logical way"—indeed, he registers a formal protest against this doctrine. He takes a sideswipe in passing at "this computer age" and disputes the idea that modern biochemistry and psychology have proved man to be "a marionette on strings, with the only difference that he was now suspended on the nucleic acid chains determining his heredity, and the conditioned-reflex chains forged by the environment." Indeed, the whole point of his book is to demonstrate that, by his scientific and artistic creativity, man escapes from the habitual action of a conditional automaton—achieving "the defeat of habit by originality."

So much for his romantic side: yet see, now, how Koestler himself describes the process by which creative originality operates. This is something for which "the workings of the nervous system" are "responsible." The scientist or artist with "the prepared mind" stops thinking about his problem, leaves his brain to get on with the job, and is eventually taken by surprise when the solution of his problem pops unannounced into his consciousness, of itself. During this Night Journey, this "inner scanning" of the mind, this retreat to free creative association, the scientist or artist is not imposing his will on the world. Rather, he is giving full rein to cause and chance, and waiting for some combination of ideas to come up which fits the precise requirements of his problem (*cf.*, the example of Franklin). Two things alone differentiate the great artist and scientist: (1) the possession of a fertile mind, well-stocked with experience, which he can leave to churn over of itself until something promising turns up; and (2) the capacity to recognise a worthwhile novelty when it presents itself, regardless of its source or origin. Where the bisociations come from does not matter, provided that he can match solutions to problems when they appear.

Yet is a theory of originality expressed in such terms necessarily anti-mechanistic at all? Leaving aside all romantic prejudices, and considering only the scientific content of Koestler's theory, one must surely answer: "No." On the contrary, the trouble with all dogmatically anti-mechanistic positions is this—the machines always catch up with them. For, in the first place, the very concept of a "machine," or of a "mechanical" mode of operation, is itself inherently vague. The machines of Descartes' and Newton's day were one thing, those of Faraday's and Edison's day another; those of our "computer age" yet a third. The thesis that "man is a machine" has, accordingly, always been inherently ambiguous— is it the *actual* machines of the time in question which are to serve as the objects of comparison, or rather some *conceivable* machines not yet built? In fact, it has rarely done scientists very much good either to assert, or to deny, this particular thesis in general terms; the only useful thing is to ask how far machines of some specified kind can be built to simulate human activity of some other specified kind.

We are therefore left with the question: "Are machines capable of original discovery?" Or rather, "Are electronic computers of a kind already in existence capable of performing those particular activities which Arthur Koestler recognises as distinctively creative and originative?" The answer to this question is: "Yes." The latest generation of electronic computers, such as those used for the programme of research on "machine-aided cognition" at the Massachusetts Institute of Technology, have in fact been built to do just that: they are capable of making inductions, of matching hypotheses to phenomena, and so of making scientific discoveries in Koestler's sense. Two things about the methods by which these "originative" or "creative" thinking-machines operate serve, indeed, to reinforce Koestler's theory about the nature of originality. In the first place, they are programmed by building into them criteria and conditions which must be satisfied by any solution to the problems they are going subsequently to be set. That is, they are given an "eye" for a solution. In the second place, they are so constructed that, when a problem is fed into them through a teletype-input, they then start "bisociating" at high speed, according to a randomised pattern determined by their electronic ("neural") connections, and they continue to do so until a particular bisociation turns up satisfying the initial requirements. Nor am I merely describing a *possible* or *conceivable* machine: such things already exist, and are being used experimentally for such practical purposes as medical diagnosis.

What does this prove? For me, personally, not very much; since I myself do not think we are compelled to *deny* the possibility of computers doing original creative thought. But I do think the existence of such machines creates a problem for Mr. Koestler, for it drives a wedge between the two elements in his position—his romantic philosophy and his scientific theory. In the long run, I believe, he will be forced to make a choice between them.

Pierre Teilhard de Chardin

From time to time in the history of ideas a man appears who, for a while, comes near to defeating all criticism: not because his works are above criticism, nor because their value is universally agreed upon, but rather through sheer elusiveness. Just because they cannot place him, his critics do not know what to make of his arguments. Since each class of reader sees him in a different light, he at first evades critical categories of all kinds; and only later, when men have been able to appraise the nature of his contribution at leisure, is a balanced judgment possible.

Such a man, surely, was the late Father Pierre Teilhard de Chardin (1881–1955). Teilhard—as he is generally known—achieved great distinction as . . . well, as what exactly? How are we to classify him? The theologians among his admirers describe him as a brilliant scientist, the scientists speak of him as a great seer or prophet, while those who knew him personally recall him as a sincere, charming, simple—even saintly—man. On the other hand, his most eminent detractor dismisses him as a naturalist, having a moderate proficiency in an intellectually unexacting kind of science, marred by an unfortunate talent for intellectual self-deceit. Evidently, questions need to be raised, not just about the *merits* of Teilhard's ideas, but about their very *status*. As well as asking whether they are "good" or "bad," we need to inquire also whether they are good-or-bad *qua* zoological hypotheses, or Catholic dogmas, or philosophical maxims, or clairvoyant predictions, or what else. In some ways, indeed, the preliminary ambiguity about Chardin's intentions and qualifications is the most urgent thing to get settled: so long as it remains unresolved, we are in no position to decide by what standards he should be judged, or in what scales his work should be weighed.

About one thing, at any rate, there is no disagreement. When his treatise on the relation of man to the natural world was published posthumously—first in France in 1955 as *Le Phénomène Humain*, then in English translation in 1959 as *The Phenomenon of Man*—it was widely read, and greeted in some quarters with enthusiasm. For almost ten years now, in France, the *bien-pensants* have treated his work as a landmark against which their own ideas must be located. Meanwhile, in the U.S.A. and even in astringent England, he has won praise from men so different as Theodosius Dobzhansky, Arnold Toynbee, Julian Huxley, and Graham Greene—though, in the background, the grating voices of less reverent critics have been heard suggesting that his divine *afflatus* is in truth so much wind.

Now, however, we can no longer defer the obligation to come to terms with Teilhard's ideas. A final collection of his essays, covering thirty years of his life, has appeared under the title of *The Future of Man*. With its help we can follow the development of his personal vision of Nature and History, and tackle the slightly ungrateful task of evaluating his general position. I say "ungrateful," not because the new volume does much to modify the accepted picture of Teilhard's personality—on the contrary, his transparent sincerity is fully confirmed—but because his supporters have, in the past, dismissed radical criticisms of his views as so much scurrility or emotional invective. In the long run, however, such tenderness will do no good either to Teilhard's reputation or to truth. His ideas will survive as something more than a fad only if doubts and difficulties candidly expressed are dealt with by his standard-bearers equally candidly.

In this essay, I shall set myself a double task—first, the preliminary "geographical" task, of determining where in the world of the mind Teilhard's doctrines are to be located; and second, the judicial task, of summing up the fundamental questions to be answered before a definitive judgment of his work will be possible. The third and final task, of actually passing judgment, is one for the future. If Teilhard's views are as significant for science and theology as his supporters declare, then the queries I shall raise here will be satisfactorily answered. This side of Alice's looking-glass, however, we should avoid anticipating that future verdict.

Teilhard's whole intellectual system was erected on foundations taken over from contemporary science: the key question, therefore, is how securely it was anchored to those foundations. His own intellectual training—his *formation professionelle*—was comparatively narrow. Its two chief foci lay, respectively, in geology and palaeontology, and in French post-idealist philosophy (Le Roy, Blondel, Bergson). By contrast, in the two subjects on which his mature ideas were most dependent—Catholic theology and evolutionary biology—his standing was that of an enthusiastic amateur. This is not in itself a criticism: at many crucial transition-points in the development of ideas, amateurs have played important and distinguished parts, since a new mind, coming to the quandaries of an established subject from outside, can often see a fresh way of cutting through old entanglements. But Teilhard's amateur status does impose on us some need for caution: while he was free to speculate about the broad implications of contemporary science, it is for us to decide for ourselves whether these implications can claim a scientific warrant.

By profession, then, Teilhard was a geologist and palaeontologist, and by personal choice he was an ordained Jesuit priest. He was born and spent his early years among the hills of the Auvergne of Central France, whose extinct volcanoes had helped open the eyes of 18th-century scientists to the true antiquity of the earth. Teilhard's family was deeply religious, and he was committed from the start to a Catholic education. The years during which his mind was formed were, in France, a time of intellectual stress, when the strong tensions between science and Catholicism were deeply felt. (This was the time, for instance, of Pierre Duhem's famous essay on "The Physics of a Believer.") By the 1890's, the first dust of the controversy over Darwinism was already beginning to subside in England; among French intellectuals, where the frontiers between science and religion have always been drawn rather differently, the full force of evolutionary ideas was only just making itself felt. While religious-minded scientists like Duhem kept skepticism at bay on the physics front, Teilhard grew up to face a debate about the new historical cosmology. This he felt—rightly enough—demanded radical re-thinking from any historically based religion such as Christianity, and the resulting questions stayed with him to his death.

From his ordination on, Teilhard pursued an orthodox scientific career as a geologist and palaeontologist, while remaining personally under spiritual discipline. After serving as a stretcher-bearer in the First World War, he returned to Paris to teach geology at the Catholic Institute, and visited China for a year in 1923, exploring the deserts along the Mongolian border. Yet already the two halves of his life—professional and personal—were threatening to conflict. In his lectures at Paris, he had put forward an evolutionary interpretation of the doctrine of original sin which his superiors now condemned as unorthodox, and his license to teach was withdrawn. He returned to China in 1926 and remained there for twenty years, partly as scientific adviser to the Chinese Geological Survey, partly engaged in palaeontological research. After the Second World War, he returned once more to France, where he was encouraged to stand for the Chair at the *Collège de France* vacated by the Abbé Breuil; but once again his religious superiors stood in his way, and he was forbidden either to apply for this position or to publish his philosophical writings.

Eventually he moved to New York, and spent the last four years of his life working under the auspices of the Wenner-Gren Foundation; so long as he lived, fidelity to his vows prevented him from releasing his writings for publication. Throughout these last postwar years he seized every opportunity to discuss the present state of evolutionary zoology with men of such authority as George Gaylord Simpson and Julian Huxley. These discussions led to significant changes in the argument of *The Phenomenon of Man*. His lifelong aim remained—to present an evolutionary picture of cosmic development capable of being harmonized with the essential message of Christianity; but, as he now saw, the way in which this was to be done had to be altered. There was one other conviction that he never abandoned. "*The Phenomenon of Man*," he declared, "must be read not as a work on metaphysics, still less as a sort of theological essay, but purely and simply as *a scientific treatise*." Here is the mark of Teilhard's fundamental simplicity. He had spent his life reflecting single-mindedly on the truths of natural science, as he understood them; and he could not believe that the outcome of these reflections might be anything else besides "science." As a sidelight on Teilhard the man, this fact is engaging, even touching. Intellectually, it cannot help being a stumbling-block.

The question—"Can *The Phenomenon of Man* legitimately be regarded as 'science'?"—is in fact the crucial one. By refusing to face it squarely, Teilhard exposed himself to the charge of wilful self-deceit which his severer critics have brought against him. The point is no mere terminological one: what is at issue is whether anything in our established knowledge of nature either compels, or could compel, us to prefer Teilhard's vision of the world to any alternative. At this point, we must try and face that central issue squarely on his behalf.

Leaving aside for the moment all the baroque terminology—"noosphere," "hominization," and the rest—in which Teilhard expressed his view of the world, we can isolate one central doctrine on which the whole intellectual construction depends. As he saw it, the historical development of the cosmos displays a rational order. Surveying the past history of the world, he finds Nature taking on in succession higher and higher forms of organization: the physical hierarchy of increasing complexity (leading from atoms through molecules and polypeptides to cells, organisms, and human beings) has been re-enacted in the successive phases of cosmic development. A creative activity or energy has been at work in the world—Teilhard concludes—of which individual human beings are the highest achievement to date. Nor is there any reason to suppose that our individual personalities are the ultimate products of this creative energy. On the contrary: we can already begin to see the emergence of a "hyper-personal" order, through which the personal wills of individual beings will be merged into a larger spiritual unity. The glory of contemporary science is that, by enabling us to reconstruct for ourselves the past phases of this evolutionary process, it helps us to plot the general direction in which the cosmos can be expected to develop in the future.

This vision was, without doubt, an appealing one. Yet Teilhard's very insistence on the scientific status of his work compels us all the more to press the questions: (a) Is this theory itself "science"?; (b) Was it in fact based on sound science?; and (c) Could it, at any rate, be reconciled with sound science?

(a) At once, a distinction must be made. Natural scientists do, indeed, concern themselves with historical developments of several different kinds—some of them involving processes explicable entirely in physico-chemical terms, others again taking place on a physiological or sociological level. On our own earth, living organisms certainly appear to have "emerged" gradually out of a pre-biotic environment; while man and the higher animals, with their capacities for thought and social life, "emerged" subsequently. Furthermore, scientists have made it their business to discover in general terms the theoretical principles exemplified in processes going on at each of these three levels; and this has led them to recognize that, at each level, the existing forms are the temporary products of historical change. So natural science has introduced us to a vast multiplicity of historical changes going on in the countless members of countless galaxies and in the profusion of living species on our own planet.

That much, however, gets us only a small part of the way toward the conclusion which Teilhard wishes to reach. If, from the unbounded variety of these historical processes, we choose to select one particular strand as representing the essential drama of the cosmos—namely, that strand of which we ourselves are the present end-product—then in doing so we are going beyond anything which science, by itself, gives us any warrant to do. What science can do is to establish the validity of certain *universal concepts*, which can be applied to all the multiplicity of processes going on in Nature—for example, the concept of historical development. But a "universal" concept is one thing: a truth about "the universe" is another. The proposition, "Everything is caught up in a process of historical change," must be carefully distinguished from the proposition, "There is a single, all-embracing process of historical change in which everything is caught up." (The statement "Every boy loves some girl" does not mean the same as "There is some *one* girl who is loved by every boy.") At the very heart of Teilhard's view lies a belief that natural science *by itself* can reveal to us the Main Road of Cosmic History, leading unambiguously from a pre-biotic cosmos up to the present state of humanity, and pointing firmly onward to a "hyper-personal" future. The existence of this Main Road, however, was something of which Teilhard was at first convinced for religious, rather than scientific, reasons. We ourselves are not compelled to follow him in this belief, unless the scientific evidence is open to this one religious interpretation alone.

That brings us to question (b): How far did Teilhard build up his vision of nature on the basis of a sound knowledge of natural science? Here the publication of his earlier essays, in *The Future of Man*, puts us in a position to confirm something which until now could be only a suspicion. At the early and formative stage, Teilhard's ideas about the origin and development of organic species owed more to Bergson than to Darwin. Consider, for instance, the following passage taken from his "Note on Progress" (1920):

> We are compelled to wonder whether the true fundamental impulse underlying the growth of animal forces has not been the need to know and to think; and whether, when this overriding impulse eventually found its outlet in the human species, the effect was not to produce an abrupt diminution of "vital pressure" in the other branches of the Tree of Life. This would explain the fact that "Evolving Life," from the end of the Tertiary Period, has been confined to the little group of higher primates. We know of many forms that have disappeared since the Oligocence, but of no genuinely new species other than the anthropoids. This again may be explained by the extreme brevity of the Miocene as compared with other geological periods. But does it not lead us to surmise that the "phyla" possessing higher psychic attributes have absorbed all the forces at Life's disposal?

To begin with, Teilhard evidently took the activity of a creative "vital force," directing the course of biological evolution toward the appearance of man, as an acceptable scientific hypothesis. This belief was in fact a misconception. No "halt in terrestrial evolution," such as he took for granted, no "diminution of vital pressure in branches of life other than the anthropoids," has any support in the findings of evolutionary zoology and palaeontology. Genetic variation in the populations of different species on the earth, and the action of natural selection on the populations—the principal processes in the evolution of species—have continued since the appearance of the first anthropoids, so far as all the evidence goes, in precisely the same way as before. The "orthogenetic" doctrines on which Teilhard originally based his religious interpretation of evolution are, by now, generally discredited.

After 1945, Teilhard at last came to accept many of the essentials of the contemporary Darwinian view of evolution:

> One must frankly admit that even at the human level life only advances hesitatingly under the influence of great numbers and the play of chance; and if one be spiritually minded one must still recognize that the work of creation presents itself to our experience as a *process* whose laws it is the business of science to investigate, leaving to philosophy the task of discerning in the phenomenon the place and influence of an *intention*.

Unfortunately, his belief in vital "pressures" or "forces" had in fact played a crucial role in the formation of his intellectual system. It was precisely the belief that at a certain point in evolution these forces had been concentrated, first upon "the little group of higher primates," then upon the anthropoid apes, and finally upon the human soul, that justified him in selecting that particular strand in evolution as representing his Main Road. Failing such an objective criterion for identifying the "direction" along which the creative pressures of "the forces at Life's disposal" are concentrated, what remains—scientifically speaking—of Teilhard's original vision? Without such a criterion is there room even for philosophers to speak of evolution as having an "intention"? If the change from Bergson to Darwin is carried right through, the picture of the totality of cosmic history as a single, directed drama falls to pieces. And, once the appearance of unity and directedness in the *past* course of evolution has been dispelled, any projection of the "evolutionary process" into the future becomes purely visionary.

The publication of Teilhard's remaining essays in *The Future of Man* thus has one helpful effect. It draws back the veil of words in which his final conclusions were clothed far enough for us to recognize the ancestry and status of his theories. However much Teilhard tried during his later years to come to terms with the Darwinian and Mendelian tradition, he never in fact succeeded in cutting himself off from his earlier philosophical roots. These roots lay, not in Darwinism at all, but in an older continental tradition, which leads back through Lamarck to the early idealist philosophers, notably Herder.

For authors within this tradition—represented in Britain, not by the professional zoologists, but rather by the philosophers Herbert Spencer and Lloyd Morgan—the word "evolution" was always the name of a cosmic process with a unique, recognizable direction and providential overtones. Throughout the early editions of *The Origin of Species*, by contrast, Darwin avoided using the word "evolution" entirely. This was no accident: his object was to produce a general theory explaining a large number of distinct and separate natural processes, not to reveal the essential plot in a cosmic drama. The dramatic interpretations of organic change put forward by Darwin's more enthusiastic followers imported into Darwinism ideas which had originated elsewhere—in Lamarck or Herder—and which, in the last resort, were probably incompatible with the Darwinian theory. In France, Darwin's explanation of the origin of species for many years made no headway; and even when the doctrine of *transformisme* was grudgingly accepted, toward the end of the 19th century, his key idea of natural selection was played down in favor of a more Lamarckian view. Even now, indeed, eminent English biologists are known to declare privately that "the French have never understood Darwin"; and there is more to this declaration than scientific chauvinism alone. Certainly it can be said quite fairly that the French have always been more interested in the philosophy of evolution than in the humdrum working-out of Darwinian natural history, and these philosophical preoccupations have tempted them to assume that some overall direction—some "intention of Evolution"—would in due course become apparent in the historical processes of organic change.

Nor was it only idealist and post-idealist philosophers who were open to this temptation. George Bernard Shaw, for instance, denounced the theory of natural selection in his preface to *Back to Methuselah*.

If it could be proved that the whole universe had been produced by such selection, only fools and rascals could bear to live.

Like all Marxists, Shaw was committed to a directional view of history, in which forms of economic organization succeeded one another in a necessary and predictable order; and such a view of human history dovetailed more comfortably with a Lamarckian account of natural history than with a Darwinian one. (The ideological support given to Michurin and Lysenko in the USSR during recent years was no mere product of temporary political factors: Russian ideologues were right to see that the intellectual affiliations of Marxism link it back to Lamarck and Herder, and differentiate it sharply from the views of Darwin and Mendel.)

The harsh truth is this: so long as genetic variation and natural selection are accepted as the dominant factors in evolution, the zoological processes by which the forms of organic species change can be interpreted only arbitrarily as displaying evidence of "intention." As Teilhard's theological critic, the Dominican Father Olivier Rabut, puts it:

> If this type of explanation is valid, there is no such thing as evolutionary gravity, properly speaking. There is a kind of statistical law of accumulation of effects; this leads, certainly, to a general trend in living forms; but no force, physical or psychical, to be compared with that of gravity; no force extending throughout evolution as a whole does in fact supervene.

There is no pre-conditioning of the variations in organic form to the new environments with which these variants will have to cope—no genetic foresight or clairvoyance. In a word, there is no *orthogenesis*. Variation goes on continuously: the superior adaptedness of certain variants shows itself after the event. The Darwinian notion of natural selection thus broke, by implication, with the "directed" notion of evolution advocated by Lamarck and Spencer.

This implication was not lost on Darwin's first readers. His own teacher at Cambridge, Adam Sedgwick, on receiving his complimentary copy of *The Origin*, protested at once:

> 'Tis the crown and glory of organic science that it *does*, through *final cause*, link material to moral. . . . You have ignored this link; and, if I do not mistake your meaning, you have done your best in one or two pregnant cases to break it.

Once the Darwinian view was accepted, it would no longer be possible to quote evidence of past zoological change in support of edifying conclusions about the moral character of the natural order.

So Darwinian theory drove one more nail into the coffin of that natural theology which had been so popular among Protestant scientists and divines throughout the 18th and early 19th centuries.

Teilhard's theories in *The Phenomenon of Man* are, accordingly, not scientific theories, but rather the current installments in a quite different tradition—the tradition of "natural theology." From the year 1700 on, religious-minded men in the Protestant world—especially in Britain—had always hoped and expected that the new science would eventually confirm and reinforce the fundamental doctrines of Christianity; and they were correspondingly ready to see in their observations of Nature evidences of "wisdom," "foresight," and "design." Those few of their countrymen, such as David Hume, who warned them against relying too much on seemingly inexplicable marvels in a natural world were ignored, or dismissed as rank atheists. All the hitherto unsolved problems of geology, astronomy, physiology, and natural history were presented as demonstrating that the world of Nature had been created as we now find it "by the Counsel of an intelligent Agent" (the words are Newton's). The result of this enthusiasm for the theological argument from design was to give a hundred hostages to fortune; and, as the physical and biological sciences succeeded in explaining the supposedly supernatural *inexplicabilia*, all of these hostages in turn had to be ransomed, one after another.

In Roman Catholic countries, scientists and theologians were more circumspect. Catholic theology never had the same need to look for direct evidences of Divine activity in the world of Nature, since it was more firmly rooted than Protestantism in Revelation and Church Tradition. So Darwin's theory of natural selection, though slow to be accepted by orthodox Catholic theologians, nevertheless escaped among Catholics the extremes of hostility it experienced among Protestant fundamentalists. From the Catholic point of view, it was not at all necessary that the story of Man's Fall and Redemption should find any analogue or parallel in the structure and history of the natural world. On the contrary: Christianity (for Catholics) had always been a religion of prophecy not prediction, built around a story of supernatural interventions in the affairs of humanity, quite discontinuous with the ordinary course of Nature.

Against this background, one can understand why Teilhard's spiritual directors found his philosophy not just unorthodox but nearly heretical. He treated the Incarnation of Christ not as the incursion into the world of a supernatural agency wholly distinct from the forces of Nature, but as one particular aspect of the evolutionary process itself. To Teilhard, the "cosmic Christ" represented "a consummation of the evolutionary process." "Christogenesis" was the final stage of the same historical process that began with "geogenesis" and continued with "biogenesis" and "noogenesis." The advance of Nature from an entirely inanimate condition, through an era of lower organisms, to our present human era—and onward, presumably, to a hyperpersonal level—was for him a process of progressive "Christification," manifesting the "spiritual energy" of "the Incarnate Christ *Evoluteur*." Father Rabut comments, slightly tartly:

> The spiritual perfection of mankind is not the automatic result of Evolution. Not only does this perfection come about through an ultra-cosmic force; not only does it depend on a divine gift, of another order from that of creation; but this supernatural gift, although it may become incarnate in the things of nature, does not do so in such a way that evolutionary progress *necessarily* leads to an increase in sanctity.

Even an agnostic like myself can see the force of this objection. As much an amateur in theology as he was in evolutionary theory, Teilhard at this point came near to adopting the position of Spinoza, for whom God and Nature were one and the same.

Taken in isolation, detached from the charm and sincerity of Teilhard's personality—which so impressed all who knew him—there is, surely, something reckless about Teilhard's theology: it reminds one of the Parson in Stella Gibbons's *Christmas at Cold Comfort Farm*, with his cheerful greeting, "Energy save you." Teilhard's description of the "direction of evolution" as "Christification" has little in common with traditional Catholicism, or indeed with theology of any kind. Such an extreme reinterpretation of Christianity drains it, indeed, so completely of its traditional message that it becomes (terminology apart) scarcely distinguishable from Scientific Humanism. No wonder Julian Huxley found so much harmony between this "revised version" of Christianity and his own secular gospel. Discussing Teilhard's vogue in private among themselves, Catholic theologians must be tempted to comment: "*C'est magnifique, mais ce n'est pas le Christianisme.*"

All the same, we can hardly leave the matter at this point. Teilhard's *Phenomenon* is not science, nor is it good Catholic theology, as science and theology are now understood. Yet is it perhaps the shape of things to come? Do Teilhard's ideas perhaps—as his admirer Canon Raven implies—represent the intellectual synthesis toward which Science, Humanism, Christianity, and other religions are alike moving? (During his time in China, Teilhard developed a great respect and sympathy for the Oriental religions.) No one embraces a heresy just for the sake of the heretic. Teilhard may have been single-minded, charming, honest, and deeply sincere; yet his personality alone would not have won him support unless his ideas had something more to offer. And clearly they had. Again and again, one finds people saying such things as "He spoke to our age." The truth is (I believe) that Christian theology grew up in association with quite different intellectual presuppositions from those we can accept at the present time. The expansion in the time-scale of cosmic history since A.D. 1750, from 6,000 years to 6,000 million, and the consequent replacement of a static picture of nature by an evolutionary one, have removed the intellectual scenery against which the Christian drama had hitherto been viewed. Teilhard's *Phenomenon* is one further attempt in a long series to re-establish a place for natural theology within the new, evolutionary view of nature.

Unfortunately for Teilhard, the full impact of Darwinism goes further than he ever recognized. Evolutionary biology at present holds out no guarantee that there exists any Main Road of Cosmic History, still less that we are on it. Nor does it present us with any "cosmic process" which can be properly interpreted as demonstrating the "intention" of the "Incarnate Christ *Evoluteur.*" Nor have we, for that matter, any reason at present to suppose that life as we know it on this earth is the only form of life existing anywhere in the cosmos. All this may conceivably change, if Teilhard's vision has the influence on science that Raven—for example— expects. For the moment, however, Teilhard's picture of man as the cosmic standard-bearer, who alone can fulfill the purposes of evolution, rests less on the established results of science than it does on human pride.

In essence, the point at issue was already an old one, before the 19th-century revolutions in geology and zoology even began. As early as the 16th century Montaigne had faced the question whether men were entitled to take an anthropocentric view of nature. Given the likelihood that other planets and satellites were inhabited by living creatures, like ourselves, what right had we men to suppose that the whole functioning of the world was arranged for our convenience?

> By what authority does Man assume that this admirable moving of Heaven's vault, the eternal light of these lamps burning so proudly over his head . . . were established and continued so many ages for his commodity or service?

Half a century later Montaigne's questions were being answered by René Descartes. There was no such authority, he replied; nor was it reasonable to demand of science that it should lend any support to an anthropocentric view of the world:

> It is not at all probable that all things have been made for us in such a way that God had no other end in view in making them. . . . We cannot doubt that there is an infinity of things which exist now in the world, or which formerly existed and have now ceased to be, which have never been seen by any man or have been of use to any.

It is an excellent thing that men should think deeply about their place in the world of nature, and relate their goals and ideals to the processes—and potentialities—of Nature. But any attempt to reverse the step taken by Montaigne and Descartes, or to find a single, unambiguous intention informing the whole course of cosmic history, must be regarded with suspicion. There may be legitimate objections to skepticism; but they are as nothing compared with the risks involved in philosophical wish-fulfillment.

Arthur Koestler II

What do we demand of Science? Vitamin-reinforced bread and astronautical circuses; *Genesis* according to Hoyle and the *Revelations* of Teilhard the Divine; piecemeal, tentative theories about those aspects of nature that we can now bring into focus; or a bit of all three? That question must not be answered in a hurry. For all three ambitions—technological, theological, and philosophical—have been operative throughout the development of scientific thought, and its history could be written with an eye to the changing balance between them.

Certainly, technology has been the junior partner in the alliance. The mask of Francis Bacon has always concealed the face of Isaac Newton, and has been used by scientists to catch patrons for their excursions into philosophy and theology. (Recall how the newly hatched Royal Society elected as its Secretary Samuel Pepys of Charles II's Admiralty, and how today's National Science Foundation was incubated beforehand within the Office of Naval Research.) Though always a selling point, technology has, intellectually speaking, never come near to the heart of science. For that one must look rather at the other two strands: the piecemeal, tentative aim well captured in Karl Popper's formula, *Conjectures and Refutations*, of conceiving and criticizing hypothetical solutions to specific theoretical problems; and the more comprehensive, speculative aim, of fitting these theoretical concepts together into an all-embracing—and, if possible, a humanly significant—view of the world.

Originally published as "The Book of Arthur." Reprinted with permission from *The New York Review of Books*. Copyright © 1968, Nyrev, Inc.

Many tough-minded scientists (it is true) dislike having science linked with theology even more than with technology, regarding their tender-minded colleagues' effusions about Creation, or design, or freewill, as a mark of softcenteredness for which Science herself (Hagia Sophia) is in no way responsible. Still: from the time of Newton until barely a century ago, the theological aims of science were accepted as co-equal with its theoretical aims. so that "natural theology" was an institutionalized element in the scientific enterprise itself. When, for instance, the Royal Society sponsored publication of the *Bridgewater Treatises* (on "the Power, Wonder, and Goodness of God, as manifested in the Creation") nobody questioned that this was a proper function for a scientific academy. Natural theology has become wholly "disestablished" from the kingdom of Science only in the twentieth century.

Even now, there are those who deplore the new order of things. In different ways, such men as Michael Polanyi and Julian Huxley, Arthur Koestler and Teilhard de Chardin all express regret at the severance of "natural philosophy" from the discussion of *Weltanschauungen*. All of them let their intellectual imaginations roam beyond the established results of disciplined scientific research, into cloudier regions of speculation—about *Personal Knowledge* as a clue to the Divine, about a religion based on Science rather than Revelation, or about a "Christogenesis" which shall be the apotheosis of both biological and human history. The appearance of the final installment of the vast trilogy that has occupied Arthur Koestler for the last ten years, or more, is an occasion for reassessing the new situation.

Intellectual courage and imagination on this scale are, in themselves, rare and admirable; but has there not been something misguided about the whole thing? Faced with Koestler's attempt to bring the concepts of the different sciences into a synthetic unity, for the sake of a general Vision of Nature that will illuminate the Glory and Predicament of Man, we must ask; "Were there not in fact strong reasons, both sociological and intellectual, why scientific theory was divorced from natural theology in the first place— reasons that still hold?"

Before turning to the actual substance of Koestler's completed trilogy, we must recall the hurdles he has set himself to clear. First, the institutional hurdle: the chief sociological mark of twentieth-century Science is the fact that it has become a profession, and the structure of scientific institutions today simply reflects this. For they are required, as never before, to be the instruments of coherent professional disciplines. That demand, by itself, has imposed on Science a new and sharper boundary, dividing a central class of problems and hypotheses, observations and experiments—whose very specificity and close relation to experience makes them the collective concern of all the scientists involved—from a peripheral class of speculative theses about the "broader implications and tendencies" of the sciences—over which there is no hope of exercising the strict rational control expected within a scientific discipline.

Thus, working as a biologist, Julian Huxley has contributed to our understanding of evolution (small "e") in ways his professional colleagues have been able to check for themselves, by seeing how his hypotheses fare in competition with their rivals, when compared with the records of our experience. Yet Huxley has been anxious to develop also a world-view embracing all cosmic history: an optimistic cosmogony in which Evolution (capital "E") becomes the central theme of History—linking primeval "pre-biotic" slime to modern democratic society, by way of all the intervening "emergent" phases of organic descent—and which at the same time will provide the ultimate justification for our moral codes, in the form of an *Evolutionary Ethics*. At this point, the problem of rational control becomes acute. To Thomas Henry Huxley, Julian's grandfather, it was equally clear that Ethics and Evolution worked not in the same direction, but in directly opposite directions: moral action should not promote organic evolution, but should suspend and counteract its brutalities. How was one to choose between T.H.'s views and Julian's? Scientifically speaking, one couldn't; for, scientifically, there was no basis for a choice between them. In moving from questions about the specific operations of organic evolution to questions about the relevance of Evolution to Ethics, they had crossed the boundary separating the hypotheses of science itself from its speculative "implications"; and in the process the whole character of the questions at issue had changed.

The establishment of this sharper boundary has had one healthy effect: it has made scientists intellectually responsive to one another's judgments, not only about the doctrines they are ready to assert, but—more important—also over the questions about which rational judgment must be for the moment suspended. In this respect they have gone beyond Socrates, for whom maturity lay in the personal acknowledgment of all that he did not know, to the position of Cusanus. Wisdom lies, for them, in the institutionalization of ignorance.

This first hurdle by itself need not be entirely daunting. The fact that the institutionalized disciplines of Science no longer find room for natural theology leaves us free to consider the broader significance of scientific ideas and insights as individuals—though it demands that we do so with proper cautions and qualifications. We are not compelled, as individuals, to act like those intellectual ascetics (and they include many working scientists) for whom the prattle of implication-hunters is as the crackling of thorns under a pot. But there is a further, intellectual hurdle, whose consequences can be more serious. Unless we are absolutely scrupulous in our handling of the ideas we pick up, our discussions of the "implications" of Science can be seriously misleading, and run into cross-purposes with the scientific debate itself. There is a standard historical illustration of this point that is directly relevant to Koestler's new book.

Our contemporary physics and physiology have developed from the New Mechanical Philosophy of Descartes and Newton; their systems were attacked during the eighteenth century both by Leibniz and by Goethe. The two attacks were quite different in character. Leibniz argued—as a matter of systematic theory—against the Cartesians' determination to explain the operation of "ordered systems" by their structure, material composition, and mechanisms alone. This attempt, he declared, inevitably distracts us from the "ordering principles" according to which the continuity and patterns of action of such systems need to be understood. Cartesian physiologists, for instance, tried to explain the passions and emotions shared by men and the lower animals as a mechanical effect of the "agitated motions" of the separate "material particles" making up our bodies and brains; and similarly for all other physiological operations of animal organs and organisms. In Leibniz's view, such a theoretical program is radically defective since, in concentrating on the component "particles," it fails entirely to explain the unity and individuality of the systems—their character as "monads." (In twentieth-century jargon: Leibniz argued that the "integrative" action of ordered systems must be understood, not by considering them as material structures alone, but in terms of the functions and activities that are characteristic of the entire "monadic" systems.)

If we now look for labels to contrast Descartes' and Leibniz's methodologies, we may (if we please) call them "mechanicist" and "organicist" respectively; but, since this was a critical debate *within* natural philosophy, the labels must—emphatically—have a small "m" and "o." Leibniz was no more inclined than Descartes to place limits on the scope of mechanistic analysis. Clearly (he said) there must be brain-processes paralleling our mental experiences and activities, to whatever degree of detail we care to probe: it is just that taking all these processes individually, and explaining their mechanical operations, leave the intrinsic character of perception and thought, as activities of the entire man, unaffected and unexplained.

Goethe and Schiller attacked from a very different angle. In their eyes, Newtonian Science was not just intellectually deficient but anathema. As they saw it, Newton had represented all material things as "blind impoverished mechanisms," and had denied the "rich purposive creativity" of living, thinking, feeling, beings:

> Like the dead strokes of a pendulum-clock
> Nature, bereft of all her Divinities,
> Slavishly serves the Law of Gravitation.

Any natural science worked out on the mechanical principles followed by Descartes and Newton—even, to be truthful, by Leibniz—must be subordinated to a broader organic vision of Nature: and the scope of mechanistic analysis must be explicitly curtailed. So Goethe the Romantic Prophet preached Organicism, in direct opposition to *all* the mathematically based programs of the New Philosophers, whom he denounced as collectively committed to a Mechanicist worldview.

Evidently, there was room here for a middle way. Newtonian physicists could be attacked as enemies of creativity only so long as the philosophical and theological strands in Science were not handled separately. Merely because natural philosophers from Descartes on had concentrated on developing and testing mechanical hypotheses about the workings of nature, they were not necessarily committed—either collectively or individually—to a Mechanicist view. Nor has there been any very consistent correlation between a man's intellectual methods within Science and the theological framework in which he has, personally, interpreted the broader significance of his own ideas. (Notoriously, Faraday was a Sandemanian.) With all affection and respect, one must insist that—over this point—Goethe was sadly confused.

All this has to be said in preface to Koestler because, in *The Ghost in the Machine*, his position is once again (as in *The Sleepwalkers* and *The Act of Creation*) that of a latter-day Goethe; and, if one keeps in mind the basic duality implicit in *Faust*, many things about Koestler's arguments come into focus. For instance: it becomes clear why, in all three volumes, he gives the impression of producing not one but two or more books at the same time—books whose arguments have unfortunately not been disentangled. For this is just what he does. In each case, he leaves it to us to distinguish the *truths* he is explaining from the *Truth* he is preaching; and these are no more connected than Goethe's Organicist *Weltanschauung* was to the organicist methodology of Leibniz.

In the first volume, Koestler wrote perceptively and intelligently about the Renaissance astronomers, particularly Copernicus and Kepler; but he coupled these sensitive intellectual portraits with a grossly simplistic Message to the effect that scientists make theoretical discoveries by a blind, unreasoning Intuition—moving unwittingly through their problem-situations like sleepwalkers across a darkened room. In the second volume, he combined some ingenious psychological analogies between scientific discovery, artistic creation, and wit, with a grandiose scheme of biological and psychological "hierarchies" intended to "break the grip" of Mechanistic Determinism, and demonstrate the potential glory of man. In both cases some intriguing ideas, expressed with all

Koestler's practiced literary skill, were presented as the basis of a revelatory message which had nothing to do with the case.

His new book continues the same pattern; only this time the display of "real science" is carried to extreme lengths. Koestler sets himself to demonstrate two things—first that, along with the "glory" of his artistic and scientific creativity, man inherits also a "predicament," in the form of a tendency to paranoia and self-destruction; and secondly, that, in the light of a properly balanced view of modern biology and psychology, this "paranoid streak in man which has made such an appalling mess of our history" can be seen to originate in our genetic make-up, to be corrected only by an "adaptive mutation."

Now if there really were adequate scientific grounds either for Koestler's diagnosis, or for his prescription, our species would indeed be in a tragic (and probably irremediable) situation. So we must look carefully both at the meaning Koestler attaches to his conclusion, and at the way he uses scientific material to support it. In the event, he gets around to discussing this central diagnosis only after the first one-third of the book: the previous 220 pages comprise his preparatory survey of the natural sciences on whose testimony he intends to rely. He begins with psychology, and here he has some sound Leibnizian points to make. Whatever merit the idea of "reflexes" may have had in analyzing the operations of the autonomic nervous system, it was always a delusion to suppose that this category could be used to account for *all* human behavior—notably, linguistic and "language-mediated" behavior. True: this delusion is still active in parts of the American academic world, thanks to the influence of Watson and Skinner; though my own observation is that the Skinnerians are by now a self-isolated minority. More relevant is the fact that even Pavlov himself never supposed that the idea of "conditioned reflexes" could be extended to the higher mental functions of the central nervous system; and his successors in Russia, such as Vygotsky and Luria, have made good use of his complementary ideas about "signalling systems" and "rule-governed communication." (Curiously, Koestler makes the same mistake as the American psychologists he so dislikes, as his few scornful allusions to Pavlov indicate: he does not realize that the theory of "conditioned reflexes" was only the *first part* of

Pavlov's physiological psychology, and one that he himself never imagined was adequate to cover the higher mental functions.)

If we adopt a more acceptable psychology, Koestler goes on to argue, we shall see that the understanding of human behavior calls for the idea of *rule-governed strategies*; and he hastens to apply the same idea to organic systems of every kind. At all levels, he claims, we must see the "directive behavior" of organs and organisms as exemplifying similar "strategies." The organic world is not a congeries of mechanisms but—to use his own neologisms—a "holarchy of holons": interacting systems whose individual operations must be understood in essentially teleological terms. This brings Koestler up against the subject which is the eternal stumbling-block for ideologues of all tendencies: organic evolution. For he cannot rest comfortably with the neo-Darwinian idea that, in themselves, the genetic mutations that occur in the germ-cells of animal populations are causally and functionally unrelated to the life-situations with which the resultant mutant individuals will have to cope. Such a belief undercuts the (teleological) idea of "strategies" at the point where he most needs it—i.e., at the intersection of biochemistry and history, Somehow or other, the biological evidence must be made to yield a notion of "mutations" which is less "random" and more "adaptive."

To this end, Koestler spends much of his second hundred pages citing authentic evidence of "adaptiveness" in the behavioral interactions between evolving populations and their environments, under the impression that, in this way, he can weaken the neo-Darwinian theory. Of course, he does nothing of the sort. All that neo-Darwinists rightly deny is that molecular changes in the ova and sperm of animals are (or could conceivably be) influenced *beforehand* by the future needs of their progeny's lives. Still, it is going to be important, for Koestler's eventual thesis, to show if possible that "adaptiveness" is built into the operation of biological systems throughout history, and right down to the lowest level. Otherwise (if I get the point of his terminology) the "holons" will not form a true "holarchy." And how could this program be carried through if one conceded that the very material of organic evolution consisted of "blind, random mutations"?

So we arrive, finally, at Koestler's central topic. His account of our human "predicament" places the cause of human paranoia in an "imbalance" between different parts of the brain: specifically, in a "dissonance between the reactions of neocortex and limbic system." If we do not always act wisely and calmly, this is because our defective brains do not permit us to do so; and only an "adaptive mutation" will enable us to overcome this defect. Such a "mutation" will have to be produced at the molecular level by beneficent "biochemical engineering." If we are to avert the ultimate historical catastrophe, we must quickly develop a Pill (his capitalization) to bring about—or imitate—the necessary "mutation."

This conclusion—the climax of Koestler's argument throughout three long books—occupies only the final ten pages of *The Ghost in the Machine*, and I have read these pages several times with great care. I have to report that I can still find no real connection between Koestler's prescription for human folly and the "scientific evidence" he offers in its support. After all that has gone before, we expect, at least, a call for an "artificially-induced mutation" aimed at "improving" the functional balance between neocortex and limbic system in the brains of our offspring. Instead, Koestler calls merely for a new, superior tranquilizer or "mental stabiliser," with which we are to control our own brains, rather than change our children's. But why do we need all the preceding argument to prove what a drink or a phenobarb will demonstrate—that applied biochemistry can help to settle the nerves?

The points at issue in this book *appear* connected, only if one is committed beforehand to a teleological interpretation of organic evolution. At least three separate theses are involved, with two further sub-theses: (i) that the major evils in human history are consequences of a species-specific psychosis, rather than (as historians assume) a mixture of genuine conflicts of interest, miscalculations, and occasional neurotic fears; (ii) that this species-wide psychosis derives from an anatomical source, in the inadequate control of the (rational) neocortex over the (paranoid) limbic system; and (iii) that such an anatomical inadequacy has parallels in the other "mistakes" of evolution—e.g., in some of the arthropods. To which one must add: (iv) the assumption that the contrast between rational and paranoid behavior has direct parallels at a neuroanatomical level *at all*; and (v) the belief that, if it were in fact clinically desirable to change the balance of action between neo-cortex and limbic system, this could be done in a sufficiently discriminating manner by pharmaceutical means alone.

Each of these five theses would take much more elaborate discussion to support than Koestler has room for; and none of them could, in the present state of things, be defended in any but the most tentative of spirits. In the event, Koestler's whole package hangs together only on account of item (iii). Here is where we run up against the central issue between Koestler's teleological world-picture and the current theories of organic evolution he finds so inadequate. For the very idea of "evolution" as "succeeding" or "making mistakes" takes for granted what is by now highly questionable: that the historical succession of organic forms can be properly interpreted as the outcome of a sequence of purposive "strategies." Without such an interpretation, indeed, what sense do phrases like "evolutionary mistake" have? Yet man is the first, and to the best of our knowledge the only, species which has *understood* its evolutionary situation well enough even to *try* to change its structure and mode of life in response to the demands of that situation; and, unless the idea of "trying to adapt" makes sense, how can the ideas of "succeeding" or "failing" possibly be applied either? It is perhaps no accident that, at this crucial point in his argument, the books to which Koestler turns for testimony and support were largely written (like Gaskell's *Origin of Vertebrates* of 1908) at a time when genetics scarcely existed as a science, and its application to the history of evolutionary change was at least twenty years away.

What has gone wrong? The answer is that Koestler is here repeating Goethe's mistake. As we have seen, his selection of scientific illustrations is intentionally tendentious: it is aimed (as he declares in the Preface) not to weigh up rival hypotheses, but to undercut an "image of man" as "a conditioned reflex-automaton produced by chance mutations . . . the antiquated slot-machine model based on the naively mechanistic world-view of the nine-teenth century." This model he sees as still entrenched in scientific orthodoxy today; and we shall not achieve a balanced conception of our essential humanity unless we replace it by an alternative "holarchic" world-view, limiting the scope of mechanistic science with the help of the distinction between "organized holons" and "mere parts."

If however we insist on the distinction between Science as theoretical hypothesis and Science as the material for *Weltan-schauungen*, it becomes possible to disentangle Koestler's message from its theoretical wrappings. It is—to make its theological character explicit—a Manichean or Zoroastrian message. Evolution has left man at the mercy of two contrary urges: a constructive one that works for good—this was the topic of *The Act of Creation*—and a self-destructive principle of evil that subverts the achievements of individuals and institutions alike. These antagonistic principles unite man to the universe, having been at work throughout cosmic history; and this insight restores the "meaning" which the "blind mutations" of neo-Darwinism took away from organic evolution. Man's hope of glory is thus balanced by his looming predicament—in the form of a genetical counterpart of original sin. Whereas Teilhard preached a bland doctrine of salvation through cosmic progress, Koestler is the prophet of a harsher biological Calvinism: though he shares with Teilhard an all-embracing cosmic vision borrowed ultimately from Lamarck, he sternly insists that the final outcome of Evolution may well be Armageddon, or the Damnation of the Species.

I have good reasons for insisting that, in the last resort, Koestler's book must be considered not as science, but as theology. Scientifically, there is never sufficient reason for choosing one worldview rather than another; and there are usually good arguments for suspending judgment and declining the choice. If Koestler had written throughout in the spirit of Leibniz rather than Goethe—simply emphasizing the need to supplement "mechanicist" explanations of physiological structures by "organicist" analyses of the functions and activities of the systems concerned—things would then have been very different. The result might then have been to improve our understanding, but it would have led to no message. In that case, however, Koestler would have had to paint a different picture of contemporary science. For, after all, there is no such entrenched orthodoxy as he alleges; and he as good as admits this in an appendix, in which he excuses himself for "flogging dead horses." Now, as always, there are branches of biology and psychology—such as embryology and language-learning—whose theories depend on "integrative" concepts, just as there are others—such as molecular biology and reflexology—that are concerned with "reductionist" analyses. The kind of modest, rationally critical mileage being got *within* Science from Leibnizian concepts could be illustrated again and again from the former branches: we can, for instance, study the life-history of the cell, or the perception of colors, or the development of mental processes in children at all, only if we are prepared to accept cells, observers, and/or children as "monadic" units or agents to begin with.

Even so, an enthronement of "integrative" concepts within biological theory—i.e., a methodological victory for Leibnizian organicism over a one-sided Cartesian mechanism—would still fall short of what Koestler's message requires. For that, what is needed is a theological victory for Goethean Organicism over Mechanistic Science of all varieties, with the entire historical process governed throughout by the rival principles of good and evil. As Julian Huxley found when he tried to argue against his grandfather, we are (for all that Science can tell us) at liberty to view the cosmic-process-as-a-whole in whatever light we please.

Those who defend a true humanism in a scientific age will do well to re-read their Michel de Montaigne. All this to-do about "blind mutations" and universal "urges" would have struck Montaigne as terribly presumptuous. Who are we humans (he would ask) to project our own ambitions onto the history of the universe? Among historians of human affairs, this lesson was learned long ago: none but the most naïve Marxian and Christian historiographers still see the Hand of God, or the Dialectic, in the detailed sequence of temporal events. The same lesson will eventually get through to our twentieth-century natural theologians also. For there is no reason to see the history of Nature, either, as pregnant with a message; and it is human vanity, equally, to claim a basis in science for an unquenchable evolutionary optimism like Teilhard's or Huxley's, or for a dark and romantic irrationalism.

P.S. After preparing this review I received a letter from Koestler's publisher:

Dear Reviewer:

On February 26 Macmillan will publish *The Ghost in the Machine*, Arthur Koestler's first new book since the *Act of Creation* in 1964.

In this book Mr. Koestler proposes a pharmaceutical solution to man's self-destructive urge: a pill to correct the streak of paranoia inherent in a man which, in this post-Hiroshima age, must inevitably lead to extermination.

The birth control pill can save man from outbreeding himself. The pill which Arthur Koestler forsees can save man from genocide.

If there were such a pill—a "peace" pill—would you take it?

Publicity Director

Above the signature was glued a red, heart-shaped pellet. I have three comments: (1) Can an author of Koestler's experience *not know* what his publishers are doing in his name? This sort of publicity speaks for itself, since it uses the language of quackery, not of science. (2) The claims made for this imaginary pill are oddly like those Leary makes for LSD, and the arguments against them are the same—that a pharmaceutical millennium would only threaten our capacity for moral and intellectual judgment, and accelerate our return to the world of *Darkness at Noon*. (3) Finally: if there were such a pill, would Arthur Koestler take it?

Jacques Monod

The rhetorical, disputatious, ideological cultural life in France (as Mary McCarthy recently reminded us) obeys different rules from those in *les pays anglophones*. To Jean-François Revel, exaggeration can even become "an artistic form" in itself; yet this fact is concealed by the belief that French is the supremely logical language. So, however ideological their aims, intellectual debates in France always affect a strict Cartesian form, according to the demands of the *esprit géométrique*. The Priest and the Schoolmaster who symbolize French intellectual life still assault each other with syllogisms. As a result, Parisian best-sellers are particularly likely to be misunderstood by Anglo-Saxons, who read their debating points as assertions, their exaggerations as dogmas, and their oratory as rigorous proof.

To make matters worse, intellectual debate in France has always been exceptionally resistant to outside ideas. Voltaire could not get anyone in Paris to take Newton's *Principia* seriously until half a century after its original publication, while Darwin's theory of variation and natural selection has taken even longer to gain acceptance. By 1859, in fact, French *bien-pensants* scientists and philosophers were already hostile to evolutionary ideas. The term *évolution* implied to them the unidirectional, progressivist view of cosmic development found in Herder and Lamarck: a historicist conception, belonging as much to natural theology as to natural science, and embracing the whole development of the universe—

Originally published as "French Toast." Reprinted with permission from *The New York Review of Books*. Copyright © 1971, Nyrev, Inc.

prebiotic, organic, and social—in a single, sweeping trajectory. Throughout the subsequent hundred years, most French intellectuals have continued to understand the term in this wholly non-Darwinian sense, seeing as the most representative ''evolutionary'' thinkers such men as Herbert Spencer and Friedrich Engels, Teilhard de Chardin and Roger Garaudy.

Jacques Monod's book *Le Hasard et la Nécessité* has to be read against this background. Monod, who has been the leading cellular biochemist at the Institut Pasteur since 1954, gave an inaugural lecture at the College de France in November, 1967, which was all that such a lecture should be: a striking call for the reintegration of fundamental biological theory with a purified ''natural philosophy,'' and for its application to human affairs. While philosophers were still fiddling away at out-of-date problems, he argued, intellectual changes were going on under their noses, notably in biology, which should be leading them to reformulate their very questions in a new frame. In particular, he called for a revival of interest in the mechanisms of intellectual history, which should lead to ''a natural history of the selection of ideas,'' and so make the evolution of human culture as intelligible, in its own way, as Darwinism made organic speciation and evolution.[1]

Coming from one of the first generation of French biologists to have fully grasped Darwinian ideas, this appeal was highly attractive, especially to those philosophers of science who had already been moving in the same direction for other reasons. Surely, philosophers cannot seriously tackle today the epistemological problems about the formation of concepts first posed by John Locke nearly 300 years ago, without paying attention to the work of Vygotsky and Piaget; nor can they hope to see a sure way past Descartes' mind-matter dichotomy while ignoring twentieth-century changes in the physics of matter, and the new insights into ''higher cortical functions'' achieved by Luria and Pribram. So one can only applaud Monod's declaration that ''one of the most urgent duties of scientists and philosophers'' was ''to contribute to a

---

1. This inaugural lecture has been published in English translation as an Occasional Paper of the Salk Institute of Biology, La Jolla.

reunification of their two fields.''[2] It is time to abandon the seven-teenth-century picture of man as separate from nature, mind as separate from matter, and to locate the problems of epistemology once again within a unified picture of the physical, animal, and human worlds.

Monod's full-scale restatement of his views has now appeared, in English translation, as *Chance and Necessity*. It has sold more than 200,000 copies in the original French; it is declared by the publishers to be "a profoundly radical philosophical statement," with implications "comparable to Einstein's challenge to Newton-ian physics"; it claims that the example of science alone can cure the *mal de l'âme* that is sapping the social and cultural life of our time. Yet Monod's most sympathetic English-speaking readers— above all, his prospective philosophical allies—can regard it only as a debacle.

Chance and Necessity is a debacle as it affects the cause that Monod himself regards as so urgent and pressing, the reunification of natural science and serious philosophical analysis into a new, biologically aware "natural philosophy." For the parochial char-acter of Monod's own intellectual milieu has trapped him into denouncing all philosophy and all philosophers alike, in terms that were scarcely pardonable even when expressed in French declama-tory prose, and that now acquire for English readers a tone of bland arrogance that only serves to expose Monod's ignorance both of the history of philosophy and of the character of philosophical issues themselves.

Certainly the scientifically minded philosophers whose good will Monod most needs to enlist, if the implications of contempo-rary biological thought are going to make themselves felt within philosophy, will only be put off by the tone of this book. They will have difficulty in sorting out the important biological points that can properly contribute to contemporary philosophy, at least as "philosophy" is understood outside Paris, from the mass of irrele-vant anti-philosophical rhetoric. If this is how one of the most liberal-minded molecular biologists finds it natural to write about

2. On this, see also Monod's address to the British Society for Social Respon-sibility in Science, November 26—28, 1970, in *The Social Impact of Modern Biology*, edited by Watson Fuller (London: Routledge, 1971).

philosophy, one can only reflect, then the Alexandrian fragmentation of intellectual life, which Monod so deplores, may well be irreversible.

Monod has three main arguments.

1. Darwinism is incompatible with any theological or progressivist interpretation of the origin of species. The basic mechanisms by which novelties—mutations and the like—enter an organic population (the "Chance" of his title) are entirely independent of the factors in its environment and manner of life ("Necessity") that selectively ensure the survival of the best adapted forms within the population. Properly understood, that discovery undermines all "unidirectional" views of natural history and development, as well as the ideologies built upon them. The deeper significance of molecular biochemistry lies precisely in this: that, having elucidated the exact mechanisms by which the inherited characteristics of organisms are both transmitted genetically and expressed in the growth of the organism, it has placed that fundamental Darwinist insight beyond doubt. So contemporary biologists have finally discredited the "cosmic animism" of traditional natural theology as well as the "vitalist" belief that the development of the physiological structure of organisms involves the operation of "non-material" agencies.

2. Ever since Galileo, scientists have weaned themselves from the belief that nature is *projectif*[3] and have committed themselves to the search for "true knowledge," which comes only from "the systematic confrontation of logic with experience." Meanwhile, according to Monod, all philosophers—aside from one or two recent existentialists, notably Camus—have clung to "animistic" and "vitalistic" presuppositions, in order to justify "various mythical histories or philosophical ontogenies" as the support for preconceived value systems:

> Ever since its birth in the Ionian Islands almost three thousand years ago, Western philosophy has been divided between two seemingly opposed attitudes. . . . From Plato to

---

3. This word causes understandable difficulty to Monod's translator, who ends by rendering his phrase, *objets doués d'un projet*, as "objects endowed with a

Whitehead and from Heraclitus to Hegel and Marx, it is clear that metaphysical epistemologies were always closely bound up with their authors' ethical and political biases. These ideological edifices, represented as self-evident to reason, were actually a *posteriori* constructions designed to justify preconceived ethico-political theories. . . .

The evolutionary biologist has at last undercut this "old covenant," linking nature and value; and has made it necessary for scientists to construct a "new covenant," in which science will take over from philosophy and teach men to live by values free of any cosmic sanction.

3. Not only has science "blasted" traditional value systems "at the root." More important, we can hope to cure the *mal du siècle*—the "profound ache" which afflicts the "modern soul" in its distress at the breakdown of the "old covenant"—only by constructing for ourselves a new system of ethical and political beliefs fashioned on the "objective" methodology of natural science itself.

So far as the biochemistry takes us, Monod's arguments for his first thesis are cogent and beautifully presented. In this respect, his book is first-rate popularization: explaining, in terms the interested layman will easily grasp, how the genetically transmitted "variations" of organic life are represented on the cellular level in the "coded" structures of the nucleic acids, and how the production and activities of specific proteins underlie the functional characteristics of the developed organism.

Over the last fifteen to twenty years molecular biologists have unraveled in great detail the roles of these two groups of substances in the cellular and physiological economy of all organisms; and the outcome (as Monod rightly feels) is among the great intellectual

---

purpose or project." Significantly, neither "purposive" nor "functional" will bear all the weight that Monod places on the term *projectif*: for the good reason that he is using the term to revive an intellectual confusion that Ernst Mayr dispelled for most biologists, with his distinction between "teleonomy" and "teleology." Thus, in true French scholastic fashion, Monod concludes his first chapter by presenting as "a flagrant epistemological contradiction" a set of paradoxes that are created entirely by the ambiguities built into his own use of the terms *projet* and *téléonomie*.

achievements of our generation. Further, its immediate consequences are just what Monod declares. If our new understanding of these mechanisms is as well-founded as it seems, no process of "directed mutation" (so-called "orthogenesis") can be conceived by which the genetic material transmitted from one generation in a population to the next is capable of reflecting and responding to the current ecological demands of an environment. Organic variation is *in all cases* the effect of causes unrelated to environmental demands. Later populations end up being "better adapted" than earlier ones only through the selective action of environmental factors.

This argument is very powerful. The success of the new molecular biology faces the few professional biologists who still support orthogenesis with grave new obstacles, and increases the burden of proof they have to overcome. But even at this early stage there is already a touch of exaggeration in Monod's presentation that puts one on guard.

One's hesitations arise at three separate points. To begin with, Monod makes excessive play with the "chance" character of organic variation, which he contrasts with the "necessity" of the subsequent natural selection. The mechanisms of variation that produce mutations are, he declares, "random," "fortuitous," and "blind"; the processes of selection by which some mutations adapt and others do not are "implacable," "demanding," and "certain." Yet while Monod's discussion of the differences between variation and selection is in many respects subtle and discriminating—he does not make the common mistake of assuming that the *explicability* of evolutionary change implies its *predictability*, but insists on the importance of unpredictable historical contingencies within organic evolution—the scientific case he has made out neither demands, nor in fact justifies, the use of this kind of language.

What neo-Darwinism and biochemistry between them make almost certain is that the mechanisms of variation are entirely "decoupled" from (i.e., causally independent of) the processes of selection. It is this decoupling that demolishes the case for orthogenesis, rather than any supposed element of "randomness" in variation itself. If we accept only this causal independence, together with the early isolation of the germ cells from environmental influence, the argument against orthogenesis will be as strong as ever, even if all mutation and recombination take place in a strictly causal—i.e., "necessary"—manner. So far as Monod's first thesis goes, therefore, there is no reason to insist, as he does, on the "chance" character of variation.

Nor is the operation of selection uniquely "necessary," either. Both in variation and selection there is room for historical contingency. An unforeseen burst of high frequency solar radiation may trigger off an unpredictable (and so "chance") wave of mutations. But so, too, the unforeseen incursion of novel predators into a territory may subject an organic population to an equally unpredictable change in selective pressures; and who is to say that these predators may not have entered the territory concerned "by chance"?

For one who is scornful of bad philosophy, Monod is curiously sloppy in his use of the terms "chance" and "necessary," which are—to say the least—problematic terms. Indeed, I suspect that his motives for placing such great weight on the notion of "chance" are not scientific but polemical. The contemporary opponents of neo-Darwinism (e.g., Arthur Koestler) themselves attack the Darwinian account of organic variation for making it "blind" and "random": instead of coolly rejecting these attacks as emotional and even meaningless, Monod prefers to stand firm and accept them full in the chest.

In the second place, was it necessary to delay the acceptance of Monod's first thesis until the last biochemical details were worked out? More than a century ago English readers recognized that the argument of Darwin's *Origin*, if accepted, was fatal to traditional natural theology. In fact, the battle over Monod's "old covenant," which still goes on in France today, had in England been largely fought over issues in geology and paleontology, even before Darwin wrote. Nearly ten years before the *Origin*, Tennyson's *In Memoriam* was already suffused with just that "anguish" which Monod diagnoses as a phenomenon of the mid-twentieth century.[4]

4. It was in fact not philosophers but conservative scientists who fought the most bitter rearguard action to preserve the "old covenant." So Darwin's geology teacher, Adam Sedgwick, acknowledged his presentation copy of the first edition of the *Origin* with a letter deploring Darwin's implied attack on the "essential link" between the moral and material world:

Only in Paris could the *rapprochement* between the Marxist Garaudy and the followers of Teilhard still have been taken dead seriously in the 1960s: to onlookers elsewhere this was obviously a last-ditch alliance between defenders of two equally superannuated forms of Lamarckism.

Outside France, the theoretical work of Auguste Weismann on the isolation of the germ cells from environmental influence in the 1890s merely confirmed what biologists already suspected: viz., the effective decoupling of variation and selection. In this respect, molecular biology has only reinforced still further an already well established position. The real question is rather, why the work of Darwin and Weismann was for so long found unconvincing in France. Did the hesitation of French biologists to accept Darwinist ideas reflect the enduring influence of some strain of mechanistic Cartesianism? Was it Darwin's and Weismann's inability to supplement their powerful theories with detailed biochemical mechanisms that (like Newton's inability to explain the "cause" of gravity) stood in the way, until now, of their full acceptance in France?

Thirdly and most important for what follows: Monod's arguments for the first thesis must be challenged at the point where he begins to write about behavior, in particular about human behavior. Though he makes passing allusions to "the second evolution, that of culture," his account of behavioral, and even cognitive, development is for the most part as strictly genetical as his account of physiological development:

> To the extent that all the structures and performances of organisms result from the structures and activities of the proteins composing them, one must regard the total organism

---

You have ignored this link; and, if I do not mistake your meaning, you have done your best in one or two pregnant cases to break it. Were it possible, which, Thank God, it is not, to break it, humanity, in my mind, would suffer damage that might brutalize it, and sink the human race into a lower grade of degradation than any into which it has fallen since its written records tell us of its history.

For an account of this early nineteenth-century debate in natural theology, see Charles Gillispie's excellent book *Genesis and Geology* (Harvard, 1951).

as the ultimate epigenetic expression of the genetic message itself.

There is a clumsiness in Monod's phrasing here that is also present in the French original. (Does he mean, "Given that all the structures and performances are . . ."; or does he mean, "To the extent that the structures and performances are . . . ," with the crucial word *all* omitted?) So long as we consider only bodily structures, this ambiguity may be trivial.

Leaving aside the effects of mutilation, disease, and under-nourishment, we need not question that the adult form of, say, a man's liver, was in some sense "written into" the genome that he inherited at birth. Behavioral performances, however, are a very different story. Certain behavioral *capacities*, like organic struc-tures, may be regarded as "epigenetic expressions of the genetic message," i.e., expressions that derive from the genetic code an individual has inherited. But the actual *performances* which mani-fest these capacities are also affected by cultural influences, and in the higher primates these cultural factors can be of dominant importance. When the snowmonkeys of Koshima were observed discovering how to swim, walk upright, wash sweet potatoes, and so on—none of which their ancestors had apparently done—they must evidently have had the "physiological prerequisites" to do these things. But does it follow that they possessed a native, genetically inherited "natatory capacity," or that this novel be-havior was solely an "epigenetic expression of the genetic message itself"? Surely these monkeys can have had "genes for swim-ming" only in a *virtus dormitiva* sense!

The moment we begin to consider topics like human language, the distinctions between bodily structures and physiological pre-requisites, intellectual capacities and manifest performances be-come crucial. At this stage in the argument, Monod would like to incorporate Chomsky's "nativist" linguistics into his own biologi-cal philosophy; yet he does not trouble to remark that Chomsky himself, both in his John Locke Lectures and elsewhere, has denied that either Darwinism or brain physiology throws any light on the status of human language. That, of course, is the beginning of another long story. But it does help to underline the fact that Monod's preoccupation with cellular mechanisms and physiological structures leaves him vulnerable when he turns to discuss thought, reasoning, language, and all those other "performances" in which "higher mental functions" are manifested: that is to say, those topics which, from the time of Descartes on, have been the specific concern of *philosophers*.

So much for Thesis One, which Monod is highly qualified to discuss. When he turns to his attack on philosophers, he is at once out of his depth, and proceeds by a sequence of sweeping and unsupported assertions. Nowhere in his book, for instance, does he pause to ask whether his picture of *all* previous philosophy as the a posteriori rationalization of preconceived value systems has a foundation in fact. On the contrary, this unexamined assumption is his only reason for denouncing all philosophers for their "subjective commitment to the old covenant," in contrast with the "scientific objectivity" required of any future system of ethics. Yet from the centuries of philosophical writing, the only philosophers whose positions he actually discusses are Bergson, Teilhard de Chardin, and Engels. (Otherwise he takes on no opponents of greater philosophical stature than Polanyi and Koestler.) And, of all these, the only man whose doctrines he states explicitly, and criticizes more than cursorily, is . . . Friedrich Engels!

Such recklessness takes one's breath away. For how would Monod himself react to a layman who gave an equally slapdash account of biology? Certainly it would not be hard to put together an equally well-based (or ill-based) defense of the thesis, "All theoretical biology has been an a posteriori construction designed to justify preconceived metaphysico-theological theories." Monod may classify Teilhard as a philosopher, but one can equally call him a paleontologist, and so a "biologist"; Jacques Loeb never disguised the fact that his interest in physiology was motivated by a deeper concern with the problem of free will; and with two or three other similar illustrations we should have as strong a case for equating biology with theology as Monod has for equating philosophy with ideology.

At this point Monod's argument is simply broken-backed. Those who share his scornful and dismissive attitude toward all previous philosophy may stay with him the rest of the way, but this attitude will inevitably alienate even his most sympathetic allies among serious philosophers of science.[5]

In philosophy, quite as much as in theoretical biology, there has been a continuous genealogy of authentic intellectual problems, which have had nothing essentially to do with *post hoc* ideological rationalization; and only in exceptional cases (e.g., a few of the nineteenth-century German historicists) have the resulting analyses been designed, among other things, to "justify preconceived ethico-political theories." From Aristotle's *Metaphysics*, by way of Kant's *Critique of Pure Reason* to Wittgenstein's *Philosophical Investigations*, the major documents of Western philosophy present complex and serious *arguments*. Monod cheerfully dismisses these arguments as eyewash, yet they are entitled to the same intellectual respect and attention in their contexts as any of the arguments of their scientific contemporaries.

From this point on, Monod's ignorance of philosophy robs his case of all force. For everything turns on the adequacy of his characterization of "science" as distinguished from "philosophy" by its commitment to the *systematic confrontation of logic and experience*. About this "postulate of objectivity," Monod remarks wistfully:

> It is hard to understand how, in the kingdom of ideas, this one, so simple and so clear, failed to come fully through until

5. As one who studied at Cambridge with both Dirac and Wittgenstein, and so was in a position to compare the intellectual seriousness and rigor of first-rate science and equally first-rate philosophy, I can only say that Monod's judgment on philosophy is irresponsibly ignorant. So is his account of it. He writes: "Ever since its birth in the Ionian Islands almost three thousand years ago, Western philos- ophy. . . ." It would be amusing to reconstruct the lost speculations of Hermogenes of Corcyra, Isostrates of Ithaca, and the pseudo-Zeno of Zante; but does not Monod see that to dogmatize about philosophy, while confusing sixth-century *Ionia* with the ninth-century Ionian *Islands*, makes him appear ridiculous—like a nonbiolo- gist's making pronouncements about physiology while assuming that an "antigen" must be an inhibitor of gene action? This is not a plea for specialization: quite the contrary. Peter Medawar has shown that a contemporary theoretical biologist can make highly distinguished contributions to philosophy, if only he approaches philosophical problems in the same informed frame of mind as he does scientific ones.

a hundred thousand years after the emergence of Homo
Sapiens. . . .

So simple and so clear? But each of the four key words in Monod's
postulate is burdened with theoretical problems and complexities at
least as grave as those surrounding the terms ''chance'' and
''necessity''; and the history of serious, non-ideological philosophy has been largely concerned with just those problems. Thus,
when it comes to testing scientific—or, indeed, ethical—concepts
and principles, what is to count as ''experience''? In what sense, if
any, can scientists bring ''logic'' face-to-face with empirical facts?
Just what is involved in ''confronting'' human theories with the
facts about nature? And how far can that procedure really be
''systematized''?

About these crucial questions Monod says nothing at all. Instead, he simply muddles along, using his four key terms in a loose
and unexplained manner. In one place he refers to the neural
synapse as ''the primary logical [*sic*] element'' in the central
nervous system. (By ''logical,'' does he here mean ''theoretical,''
or is he comparing the synapse to a ''logic'' unit in a computer?) In
another, he hints at a physiological basis for the synthetic a priori,
such as Kant toyed with in his younger days but later abandoned.
Nowhere does he acknowledge the ambiguities that are built into
his position by the obscurity of his own basic ''postulate.''

If Monod had taken more care in reading the philosophy that he
dismisses he might have understood that the task of clarifying these
complexities in the notion of ''objectivity,'' and of refining the
procedures of scientific investigation in the light of that clarifica-
tion, has been a major preoccupation of Western philosophers since
the seventeenth century, and that, in this activity which still goes on
today, philosophers and natural scientists have been not mutually
scornful rivals but effective and necessary *allies*. The scientific
work of Helmholtz, for instance, was greatly indebted to epistemological analyses carried out by Kant half a century earlier, while
in the theories of Einstein, who knew his Hume, and Bohr, who
admired William James, the line between epistemology and theoretical physics became paper thin. In this difficult work nineteenth-
and twentieth-century French philosopher-scientists, such as
Bernard, Duhem, Poincaré, Meyerson, and Bachelard (not one of

whom Monod even mentions), have played a significant part.

By the time he comes to defend his third thesis, Monod has no solid shots left in his locker. In fact, he has inadvertently talked himself into an existentialist position. For the basic structure of his ethical arguments was already familiar many years back to Kierkegaard, Schopenhauer, and Nietzsche. (Once again he seemingly forgets that there were existentialists long before Camus, and that away from Paris the "old covenant" between value and nature has been under powerful attack for a century and a half.)

So after all the fanfares and promises, he has no way of redeeming his earlier promissory notes. In response to the "anguish" at the breakdown of the "old covenant" which he diagnoses in the "modern soul," he has nothing to offer except rhetorical phrases. A slogan like "the knowledge of ethics must be founded on the ethics of knowledge" sounds all too impressive in French, yet it does no more than the thoughts of Chairman Mao to bind up "the wounds in the modern soul." Nor does the emptiness of Monod's prescriptions give one any better confidence in his diagnosis of this *mal du siècle*. Rather, some of his passing remarks make one wonder whether, even now, the full force of the Darwinian methodology has really struck home in France.

In late nineteenth-century Britain and America the intellectual assimilation of Darwinism did much to encourage pluralism, pragmatism, and utilitarianism—especially in the social and political realm—of kinds that make Monod unhappy. In a recent discussion in London, referred to above, an Indian participant quizzed Monod about his demand for an "ultimate justifying criterion" for our "value system"; provided that one approached ethical issue always in a critical spirit, was not the search for "ultimates" a wild goose chase? Monod replied. "Well, this is a typical criticism from a typical British empiricist," and went on to insist that mankind cannot survive without a *systematic* code of values, backed up by some sort of a myth—even if one rooted in science.

Yet this reply is simply not good enough. Why do all our "values" have to be linked together into a single "system" at all—to say nothing of a system bolstered up by a "myth"? A truly Darwinian thinker would surely view the values, institutions, and social structures of a people as forming not a "system" but a "population"—a population which is more or less well adapted to the needs of the men concerned, and within which individual practices can change more or less independently in the face of new socio-historical situations, in a pragmatic manner, and with more or less "adaptive" consequences. So it is probably no accident that Monod's own exposition of evolution theory is preoccupied with biochemical mechanisms, and neglects the other equally indispensable aspect of the subject: ecology and population dynamics.

From the beginning Darwin himself thought of species and evolution in ecological and populational terms; and the construction, during the last fifty years, of an integrated neo-Darwinist biology owes quite as much to the sophisticated population dynamics of men like R. A. Fisher and Sewall Wright as it does to the molecular biologists. Indeed, if neo-Darwinism is to be seen as offering a method and an example to thinkers in other fields (e.g., ethics, sociology, and political theory), the reanalysis of such notions as "adaptation" in the light of the behavior of populations has intellectual implications at least as profound as those of the new mechanistic biochemistry.

In Paris Maurice Merleau-Ponty is dead. Jean-Paul Sartre is given over entirely to politics. Only a very few of the established philosophers (notably Michel Foucault) are clearly concerned with the same subjects as their colleagues in other countries. The Priest and the Schoolmaster have been joined by the Commissar in a triangular battle over "truth." As a result, Monod's *Le Hasard et la Necessité* has been the sensation of the season, and the object of a grand ideological debate.

Historically, perhaps, we can see the book as a return swing of the same pendulum that, fifty years ago, carried French intellectuals away from the "scientism" of Taine and Renan to the Catholic "anti-scientism" of Péguy and Maritain. Certainly, Marxists and Catholics alike have attacked Monod's views in the name of a natural theology that died long since elsewhere among all but a fringe of philosophers and biologists. Yet there is something essentially unfruitful about this ideological way of linking biology and philosophy, and all the associated brouhaha only distracts attention from Monod's initial plea: that we should look for the true points of contact between the novel insights of contemporary biology and the outstanding problems of philosophy and epistemology.

As early as the turn of the century, men like Ernst Mach and Georg Simmel, John Dewey and Charles Sanders Peirce were already taking hold of the constructive philosophical issues arising out of Darwinian theory: it was the successors of that notorious ''Social Lamarckist,'' Herbert Spencer, who got hold of the wrong end of the stick and perpetuated the pre-Darwinian debate within the social sciences. It is only Monod's personal involvement in molecular biochemistry that leads him to exaggerate its broader philosophical implications today.

For why should the results of all this splendid scientific work on the sub-cellular mechanisms of mitosis, meiosis, and so on, come as any kind of shock or surprise to Kant, or Descartes, or Aristotle? Could not all these men have accepted the idea that the exact structure of, say, an adult human liver is prefigured in the ''form'' of the embryo, without the slightest difficulty? Of course, it is a fine thing to know just *how* that prefiguring is in fact achieved and expressed: yet, even for a modern Aristotle, the new molecular biology would simply demonstrate in explicit detail the manner in which the *potentiality*, or ''genetic message,'' of an organism is *actualized*, or ''epigenetically expressed.''

Philosophers do indeed have much to learn from contemporary biology, and no one in either camp can afford attitudes of professional specialization. But almost all of these significant points of contact are in fields that Monod's book says very little about. There is, for instance, the large topic that he referred to in his inaugural lecture: the process of conceptual variation and intellectual selection responsible for the ''second evolution'' of culture and ideas. Behind this, there is a further exciting and important topic: namely, the manner in which the products of culture—language, ideas, conceptual discriminations, and the rest—are embodied and ''represented'' in the cortex, during infancy and childhood, as the physiological counterparts of computer ''software.''[6]

6. In this respect, the book *Higher Cortical Functions in Man* by A. R. Luria (Basic Books, 1966) is very suggestive: particularly in relation to the secondary layers of the sensory regions.

Then, again, there is all that we have yet to learn from studies of animal behavior: about communication, behavioral capacities, and the beginnings of cultural learning and social structure. In each of these cases, however, the part played by molecular biology may be intriguing, but its implications are not crucial. The significance of theoretical biology for philosophy—such as it is—still lies in its power to throw light on those perplexing problems that faced Descartes and Kant, and still face philosophers today: that is to say, the nature, interrelations, and embodiments of "rationality," "consciousness," "thought," "language," and the other higher cognitive functions.

François Jacob

After a century of tissue rejection, the grafting of Darwinian evolution theory into the body of French thought is at last proceeding with all deliberate speed. One says "deliberate" because even now very few true evolutionary biologists are to be found in either the universities or the medical schools of France. The paleontological and physiological traditions of Cuvier and Bernard are still influential among French academic biologists and medical scientists; and neither of these essentially structural modes of thought displays any great compatibility with Charles Darwin's *populational* approach. Yet at the pinnacle of the scientific hierarchy, at the Institut Pasteur and the Collège de France, there is a small group of remarkable biologists who are at last determined to achieve a successful transplant, and to install within the Cartesian world picture of French science a functioning account of variation, natural selection, and the rest—ideas that have hitherto provoked in France the reactions proper to "foreign bodies."

The two principal figures in this group are Jacques Monod and François Jacob, who are internationally known for their "operon" theory. This theory is designed to explain how the biochemical processes taking place along macromolecules in cell nuclei are switched on and off, and so to account for the puzzling latency and intermittence of gene action. For why are not all genes active all the time? The philosophical attitudes underlying their enterprise were first presented to a wider public a few years ago in Monod's book *Le Hasard et la Nécessite*[1] and now we also have Jacob's *La*

Originally published as "A Biology of Russian Dolls." Reprinted with permission from *The New York Review of Books*. Copyright © 1974, Nyrev, Inc.

1. See above pp. 140–155.

Logique du Vivant, in a perfectly respectable and readable translation.[2]

Jacob's book was well worth waiting for. It is an altogether more serious piece of work than Monod's, more tightly argued, better organized, and less rhetorical: an admirable addition to the popular literature of French science. Where Monod put forward a lawyer's brief, in which the crucial distinction between the intentional (teleological) activities of human beings and the mechanically self-correcting (teleonomic) processes of biology was deliberately obscured behind the common word *projet*, Jacob tells his story absolutely straight. As a result, it is possible to see at last exactly what is going on: i.e., how this improbable union of neo-Cartesian physiology and neo-Darwinian natural history is being effected.

To put the point in the kind of biological terms that Jacob himself appreciates,[3] the Institut Pasteur group sets out to achieve a viable transplant of Darwinian evolution by using the new science of molecular biology to inhibit rejection. Hitherto French biologists have found the theory of natural selection uncongenial because there was no clear way of meshing it in with the mechanistic view of bodily structures they inherited from Descartes. Until the creation of molecular biochemistry in the years after 1945, by Delbrück and Kendrew, Crick, Watson, and others,[4] there was indeed no way of relating the population genetics on which neo-Darwinism rests—or even the chromosomal genetics of T. H. Morgan and his successors—to the actual physicochemical make-up of organisms and their minute parts. Molecular biology supplied that link. By means of the new concepts of macromolecular struc-

2. There are a few easily corrected solecisms: e.g., the standard Anglo-American mathematical usage corresponding to the French word *combinatoire* (the subject of a translator's note on page 30) is not "combinative," as she suggests, but "combinatorial" or "combinatory."

3. Cf. "the fusion of cultures is like that of gametes; the university in society plays the role of the germ line in the species; ideas invade minds as viruses invade cells. . . ." (p. 321).

4. Interestingly, Jacob gives explicit historical credit for molecular biology only to Avery, its grandfather, and Schrödinger, its godfather. He leaves the actual architects of the new science anonymous, as "a school of crystallographers," "many young physicists," etc.

ture and the "genetic code," it was shown that the unit factors of heredity first invoked hypothetically by Mendel, and later established by twentieth-century geneticists, have a quite definite and specific "structural" aspect. In a manner of speaking (it seemed) genes actually "were" molecules of nucleic acid.[5]

That is the key discovery toward which François Jacob's entire exposition is directed. He is writing a history of ideas about heredity and organic reproduction, from the late sixteenth century to the present day—but it is history with a philosophical message and a ready-made destination. From 1600 on (he argues) the interlocking structures of *le vivant* have been unraveled step by step; the focus has narrowed from complete organisms to constituent organs, to cells and cell nuclei, chromosomes, genes, and finally macromolecules; and, by now, the hidden structures and mechanisms that impose a common *logique* on all the functions and operations of the living creature stand revealed. Inside every organism, smaller and smaller structures and systems are packed one within another like Russian dolls—the image is his own—and at the very heart, on the smallest scale of all, they can at last be described and explained in physicochemical terms.

Separated for so long from physics and chemistry by differences of method and concept, biology, Jacob argues, can at last join hands with the physical sciences. After a century of quarantine, Darwin can be made an honorary Frenchman:

Each living system has to be analyzed on two planes . . . which can be separated only for the sake of explanatory convenience. On the one hand, one has to distinguish the principles governing the integration of organisms, their construction, their functioning; and on the other, the principles that directed their transformations and their succession. The description of a living system requires reference to *the logic*

5. Cf. "the biologist will not rest until he has replaced it [the gene] by material components . . . as if, in order to last, a biological theory had to be based on a concrete model" (p. 14). Jacob should have written, "The *Cartesian* biologist will not rest. . . ."

of its organization, as well as to *the logic of its evolution.*
Today biology is concerned with the *algorithms* of the living
world. [Page 300, italics added]

Given his overall synthesizing aim, and the compression that
comes from dealing with 400 years in 280 pages, Jacob handles his
historical tale with grace, style, and reasonable accuracy. There are
some inevitable obscurities. For example: on page 47, he quotes
Buffon calling in question the reality of all taxa, by saying, "There
are really only individuals in Nature," yet five pages further
on—without warning—we find Buffon contradicting himself,
"An individual is nothing in the universe. . . . Species are the only
beings in Nature."[6] Still, when all allowances are made, Jacob's
historical survey is exposed to only one major objection; though
this particular criticism (as we shall see) is highly damaging to his
deeper philosophical message.

The point of contention has to do with the temporal or historical
aspect of the theory of evolution. For Darwinists, the manner in
which organic populations become adapted to their specific environments, and to one another, is quite as crucial a question as the
physicochemical nature of the mechanisms involved in heredity. In
Britain and America, accordingly, modern evolution theory has
rested on two quite distinct pillars: one of them being cellular
genetics, from which Jacob himself starts, the other being the
application of statistical and genetical analysis to the dynamics of
evolving populations. (This latter, populational analysis, was inaugurated by Sewall Wright, R. A. Fisher, and J. B. S. Haldane,
and has since been refined by Ernst Mayr and G. G. Simpson, and
more recently by Richard Lewontin, Richard Levins, and others.)

Despite all his concern with the "logic" of evolutionary
change, François Jacob—believe it or not—tells us absolutely
nothing about this second, crucial support of evolution theory.

6. This change of mind on Buffon's part was a historically important shift of
front, which Jacob should not merely have glossed over by saying that "the species
was never a source of the arguments that were provoked by the genus." If anything,
the species concept has been a source of much graver arguments than the genus
concept; for nobody could doubt that the boundaries between species had some
reality in nature.

While he acknowledges that "adaptation" is a key feature of organic evolution, he mentions no attempts to analyze the concept of a "population" more recent than those of Malthus and Darwin. True: he allows that "the study of large populations and the introduction of the statistical method" has had important consequences for biology (page 195); but this leads him to discuss only Maxwell, Gibbs, and Boltzmann, the founders of statistical thermodynamics. So far as the most eminent French evolutionists are concerned (it seems) Fisher, Wright, and Haldane have written entirely in vain!

This omission is a bit of a giveaway. For the charm of Jacob's basic analogy between structure and history—between the *logique* of physiological organization and the *logique* of temporal evolution—depends on keeping the historical messiness of populational changes under a hat. Statistical thermodynamics and information theory, its stepchild, are in fact theories of a very different kind from population dynamics and population genetics. The former are decent, ahistorical, Platonizing systems, concerned with physicochemical Being, and highly congenial to the Cartesian spirit. Population dynamics and genetics, by contrast, provide an essentially temporal, neo-Aristotelian account of organic Becoming, in which historical factors are the heart of the matter, not mere "boundary conditions" of some more basic, ahistorical theory.

If the adaptation of organic populations were simply one further statistical phenomenon governed at bottom by the laws of large numbers, like Boyle's law of gas expansion, we might see it as yet one more consequence of timeless, mechanical processes. Unfortunately for Jacob's case, this is not so. Darwinian evolution theory remains irreducibly historical, and continues even now to resist exhaustive translation into the ahistorical terminology of physics and biochemistry.

So there is something premature, even specious, about Jacob's suggestion (pp. 299–300) that with the advent of molecular biology the older dialectical stresses in biological theory—discrete species vs. continuity of forms, mechanism vs. vitalism, etc.— have finally been outgrown; and about his idea that we can now construct a unitary pattern of theory—extending all the way from macromolecules to individual organisms, and on to organic species, societies, languages, and cultures—around a common general idea of organized systems, or ''integrons.'' (The neologism is his.)

Even within biology itself, to say nothing of human affairs, the dialectic between physical structure and historical adaptation is unresolved, and probably unresolvable. At the opposite extreme from such molecular biochemists as Jacob, for instance, there are those contemporary population biologists (e.g., Lewontin) who see subcellular mechanisms, of DNA replication, etc., as quite secondary to the understanding of evolution. Far from DNA macromolecules being identical with genes, they are merely the tactical means that populations employ in responding to the strategic, and historical, problems of adaptive change in varied environments. Nor will this opposition be easily overcome. In biology as in the human sciences, structural perspectives—in the spirit of Descartes and Newton—can at most complement, never displace, the temporal perspectives of Vico or Darwin.

The root of the trouble is this. Jacob declares, rightly enough, that

> Biologists no longer study life today. They no longer attempt to define it. Instead, they investigate the *structure* of living systems, their *functions*, their *history*. [Page 299, italics added]

Yet there is no doubt that, for him, the greatest of these three is ''structure.'' Organic functions and history become truly intelligible to him only when accounted for in structural terms. Jacob's entire history of heredity is, indeed, designed to establish a structural picture of organisms and their activities; and his parallel between the ''logic of organization'' and the ''logic of evolution'' leads the reader to understand that, once this is done, the temporal sequence of organic evolution, which is biological history, will prove to have a ''logic,'' conform to a ''program,'' and be governed by a ''code'' or ''algorithm,'' analogous to the logic / program / code of macromolecular structure.[7]

7. Even Jacob's ''Russian doll'' model of the organism is question-begging. For the relationship between more or less complex physiological ''systems'' in the

This point is more than a parochial one. For the final twenty-five pages of Jacob's book range, in conclusion, over the entire field of life and culture, behavior and society. In short order, Jacob sweeps sociology and linguistics, anthropology and political theory into the net of his Biological Structuralism:

> A new hierarchy of integrons is thus set up. From family organization to modern state, from ethnic group to coalition of nations, a whole series of integrations is based on a variety of cultural, moral, social, political, economic, military and religious codes. The history of mankind is more or less the history of these integrons and the way they form and change. [Page 320]

There are essential analogies, in his view, between Nature and Nurture or between genetic inheritance and sociocultural transmission; and these analogies are to be captured by adopting the language of information theory—i.e., by analysis in terms of "codes," "messages," and "programs."

This vast extension of Jacob's natural philosophy into human affairs may be appealing to students of transformational grammar.[8] But the rest of us would do well to tread more carefully. For human affairs—notoriously, in Descartes's own view—lack precisely the kind of tight and self-correcting, or "systemic," organization typical of physiological systems in organisms. The "organization" of entire societies is something less than organic; the behavioral "codes" of human groups are only very partially codifiable; the

---

organism is not adequately depicted as one of *spatial inclusion*. Rather, these systems have very different kinds of functional complexity; and the most elaborate and sophisticated systems—e.g., certain crucial parts of the central nervous system—can be rather small.

8. Cf. "According to modern linguistics, there is a basic grammar common to all languages; this uniformity would reflect a framework imposed by heredity on the organization of the brain" (p. 322). With all respect to Jacob, the thesis that the shared "deep structures" of language are directly represented in the brain, in ways that are genetically inherited, is in fact both highly speculative and probably incorrect, on both neurological and evolutionary grounds.

so-called "social system" is a good deal less than systemic.[9] In short, the integration of Jacob's "cultural and social integrons" is far from complete. If the course of History did indeed conform to a standard mathematical procedure or "algorithm," how happy that would have made Hegel! Failing that, the extension of terms like "logic," "code," and "algorithm" from molecular biology into evolutionary history—and still more into human affairs—involves a somewhat questionable set of metaphors.

Here again, of course, Jacob's sketch of a sociology of integrons is in a familiar French tradition. (When did he last read Durkheim?) Yet the deficiencies inherent in any organic theory of politics, culture, and society are not overcome simply by restating it in the terminology of information theory and molecular biology. Far more to the point is Jacob's suggestion that we explore further the parallels between sociocultural change and organic evolution.[10] An evolutionary analysis of the processes by which cultures and institutions, languages and concepts change over time will, however, have to be an essentially historical and populational analysis. So the evolutionary analogies we need in order to improve our understanding of society and culture are those that come, not from molecular biology and cell physiology, but from population dynamics and population genetics: i.e., from precisely those aspects of organic evolution that Jacob himself ignores.

After all, syntactical and other "codes" can acquire a semantic sense only when put to use in larger human *Lebensformen*; "programming" is an activity by which computers are enculturated to *our* historical situations and problems; while—as our experience with nineteenth-century historicism testifies—"logic," even when supplemented by "dialectic," has proved a feeble instrument

9. After years of loose talk about "ecosystems," many leading ecologists are now shying away from that term, for similar reasons. The phenomena so referred to (food-chains, etc.) also lack the stable, self-restoring character of physiological systems: i.e., are not fully "systemic." If only they were!

10. "In short, the variation of societies and cultures comes to be based on evolution, like that of species" (p. 321). On this whole topic, let me refer to the writings of Donald T. Campbell, and also to my own *Human Understanding* (Princeton, 1972), especially chapters 1, 5, and 7, which discuss fully the non-systemic, populational character of sociocultural evolution.

for the analysis of historical processes.

Like all Cartesians, the structuralists will no doubt go on trying to enclose the flux of history in fixed, universal categories; and the results of their efforts will often be illuminating, up to a point. But, in the last resort, the Protean character of history—especially, the history of human life and problems—will continue to overflow all such tidy boxes.

Carl Sagan

The intellectual landmines laid for us by Charles Darwin more than a century ago continue to explode. As each cloud of dust settles back around us, we begin to see a little more clearly the ways in which the history of humanity interlocks with and reflects the continued presence of patterns apparent in the history of nature. Yet each of these novel insights in turn finds itself resisted by those who have a major investment in the *uniqueness* of the human species, and who feel themselves threatened by any new analogies between the human species and other kinds of animals—most particularly, between the modes of life and experience of human beings (their psyche and ethos) and those of other primates and higher animals. At times, the passions aroused in the resulting controversies even remind us of the outburst that greeted the original publication of *The Origin of Species* in 1859: recall, for instance, last year's angry exchanges in response to Edward Wilson's book, *Sociobiology*.

From the standpoint of the development of ideas in the long term, however, these opponents seem to have been living in something of a fool's paradise. When Darwin finally let *The Descent of Man* go to press in 1871—his argument in the *Origin* had been discreetly silent about the evolutionary status of human beings—he claimed explicitly only that "Man still bears in *his bodily frame* the indelible stamp of his lowly origin." As a result, the first scientific and philosophical battles about human evolution were all fought out around the morphological features of different species, around

Originally published as "Back to Nature." Reprinted with permission from *The New York Review of Books*. Copyright © 1977, Nyrev, Inc.

fossil bones and brainpans, the beaks of finches and the necks of giraffes. And since the scientific case for the animal ancestry of the human species had been largely made out on morphological grounds, it seemed possible for a while to resist the further encroachment of evolutionary ideas by adopting a policy of containment. Evolution had to do, not with ethos or psyche, but with *morphe* alone. The mental and moral sciences could continue to hold themselves aloof from the natural sciences. *Geist* remained distinct from *Natur*. The methodological dualism which Kant had substituted for the ontological dualism of earlier Cartesian natural philosophy still seemed to protect human history from any danger of being engulfed in natural history.

This defense line was, at best, a temporary one. As early as 1838, Darwin himself had been thinking of ways in which his vision of man as part of a single creation with all other animals could be extended into psychology also. In private notebooks that remained unpublished in the Cambridge University Library until the 1970s[1] he confessed himself a "materialist" in his views about the function of the human brain as the organ of higher mental activities. Yet he understood well enough how much damage would be done to the reception of his evolutionary theories by any public profession of these views: not least from memories of the storm provoked by William Lawrence's *Lectures on Physiology, Zoology and the Natural History of Man* (1816–1819), which was still blowing strongly against the materialist position when Darwin was a medical student at Edinburgh University in 1825–1827.[2] So it was only toward the end of his life that Darwin allowed himself to give some hints of his ideas about mental evolution and the psychic affinities between humanity and the other species: e.g., in his book, *The Expression of the Emotions in Man and Animals* (1872), a pioneering study in the comparative psychology of affect, which

1. These "M" and "N" Notebooks have been admirably transcribed and edited by Paul H. Barrett of Michigan State University and published with commentaries and a long introductory essay by Howard E. Gruber of Rutgers, in the book *Darwin on Man* (Dutton, 1974).

2. Gruber and Barrett record a curious incident at Edinburgh in March 1827, which shows how violent the feelings on this topic still were: *op. cit.*, pp. 39, 479–480.

has been followed up only in our own day, and is still not generally appreciated at its true worth.

The task of putting human nature and affairs (psychosocial as well as physiological) into a single perspective with natural science and natural history has been a slow and painful one.[3] The cosmologies and natural philosophies of classical antiquity and preclassical times saw the chronicle of human history as played out against a static backdrop of the natural world. Though Heraclitus might declare that *everything* was "in flux," only the Epicureans took very seriously the problem of relating the flux of human affairs to that of Nature herself. On this issue, Platonists and Aristotelians were for once lined up on the same side, along with all those historical writers who followed Thucydides in seeing history as the record of human character or personality displaying its characteristic virtues and weaknesses within the unfolding of political and social events.

If most classical thinkers set Humanity over against Nature, this was of course even more clearly the case with the Christian story of the Fall and Redemption. Until 1700 or later, God's Natural Creation was almost universally perceived as unchanging (at any rate within the present Divine Dispensation), and served as a fixed stage for human beings, who were alone directly and intimately involved in the historical drama of sin, grace, and salvation. So it is worth noting how much care was taken by those first few historically minded eighteenth-century speculators (e.g., Vico) for whom the origins of human society, customs, and feelings once again became a serious theoretical issue to rebut the charge that they were "Epicureans." Spokesmen for the conservative *bien pensant* position certainly knew who was the enemy, and how to blacken anyone who strayed in that direction. (More recently, this has been one of the issues that divided Hegel from Marx. For the historical idealists, the "historical dialectic" affected only *Geist*, as mani-

3. The obstacles to this task, failing a realistic sense of the vast extension of cosmic time, are one of the chief topics discussed in my own examination of the stages by which natural history and human history gradually negotiated some kind of a common framework and timescale, in *The Discovery of Time* (Harper and Row, 1965).

fested in human history, while *Natur* was perceived as essentially repetitive, not progressive.) And, even today, it cannot be said that many "human scientists"—whether anthropologists or historians, political theorists or social philosophers—view Darwin's longer term project for integrating our understanding of human nature with that of our fellow animals, in their mental and social as well as their somatic aspects, with any real relish or eagerness. Too often, the topic awakens in them only irritation or distaste. A good illustration of this is Marshall Sahlins's polemical response to E.O. Wilson's *Sociobiology*, which presses the argument for recognizing the uniqueness of the *Geisteswissenschaften* against Wilson with a certain weary impatience.[4]

Carl Sagan's engaging and well written new book, *The Dragons of Eden: Speculations on the Evolution of Human Intelligence*, is an antidote to much of the recent controversy about human evolution. His style and manner have a certain briskness and astringency, an alkalinity perhaps, which come pleasantly to the reader after the acidity of last year's narrower debate over Wilson. Sagan himself is best known as an *aficionado* of space travel, and as a man much preoccupied with the problem of establishing contact with any intelligent beings there may be in other planetary systems or galaxies; he was most recently in the public eye in connection with the Viking landings on Mars. But we find him here writing not about astronomy or interplanetary technology so much as about brain physiology and linguistics, evolution and psychology, dreaming and DNA. He does so not as a specialist in one field of science who has been tempted to wander out of his proper field and hazard amateur speculations about other scientists' subject matter. Rather, the book shows that he is a true "natural philosopher," whose concern with extraterrestrial intelligence is only one element in a larger scientific program, and whose real goal is to produce a revised version of the story of human history and destiny, within the boundary conditions set by the ideas of twentieth-century natural science.

His canvas is thus a broad one. In showing us the psychological significance of our evolutionary ancestry, he seeks to balance off

4. See *The Use and Abuse of Biology* (University of Michigan Press, 1976).

the respective contributions to human intelligence and mental life of our genes and of our brains. Looking at the sweep of evolutionary history, he concludes that there was a "gradual increase through evolutionary time of both the amount of information contained in the genetic material and the amount of information contained in the brains of organisms"; but as time went on reptilian creatures evolved which for the first time had more information capacity in their brains than in their genes. Since these early creatures—probably a few hundred million years ago—there have been two major bursts of brain evolution, involving the development of the limbic system and of the neocortex, and associated with the emergence of mammals and the advent of manlike primates. As a result, we inherit a "triune" brain whose threefold structure and modes of functioning (reflected in the metapsychologies of Western thinkers from Plato to Freud) need to be related to their evolutionary history. Only if we understand the manner in which the human brain has come to its present form, Sagan argues, can we speculate at all fruitfully about the directions in which human intelligence and its bodily agencies are capable of evolving further in the future.

The particular problems on which Carl Sagan rightly chooses to focus his discussion have to do with the evolution of the brain and its function as the organ of human feelings, experience, and intelligence. Like many other modern observers, he is struck by the evolutionary "modernity" of the neocortex, and by the complex ways in which its operations monitor, yet are constrained by, the operations of parts of the brain that are more deeply seated and "older" as far as evolution is concerned. Where some recent popular writers, such as Arthur Koestler, have treated the limbic system and other deeper structures as having crudely emotional functions, over which the somewhat frail neocortex has only marginal rational control, Sagan gives us a more subtle and historically informed view.

Those of our evolutionary ancestors that had not yet developed any major cortical systems, he argues, were adapted to alternative modes of life, from which many of our own intelligent, and essentially linguistic, operations were absent. But they did not lack control and monitoring mechanisms of their own kinds, adequate to their proper modes of existence. Moreover, to the extent that the deeper structures of our modern brains still embody the mechanisms that were formed in the crucible of that earlier existence, we too can still—in sleep, in dreams, perhaps even in the echoes awakened by our myths—get back in touch with that earlier mode of life.

The "dragons" of Sagan's title refer to the scaly winged creatures halfway between the dinosaurs and the birds that may have been the chief predators against which those earlier brain formations were our ancestors' prime defenses. The collective memory of those far off but crucial days in a geological "Eden" may still, Sagan hints—and not entirely in a spirit of whimsy—be preserved in the widespread occurrence of legends about such dragons.

This discussion of brain structure, both as a product of earlier evolution and as the instrumentality of human intelligence, is only the center piece of Sagan's larger picture of world history. Writing in the tradition of Buffon's *Epoques de la Nature* (1778), he begins with an over-all account of the "cosmic calendar," as natural scientists now conceive it. Where Buffon chose to map the successive phases in the history of the earth against the traditional Days of Creation, Sagan invites us to see cosmic history against the time scale of a single year. By that measure, he tells us, life began on earth only around "September 25," while the human species made its appearance around 10:30 P.M. on "December 31"—recorded history being represented by the last ten seconds alone. Presenting the vast abysms of unrecorded time in this kind of way can be irritating to humanists, since it may appear intended as a kind of clever putdown. In Sagan's hands the impression is different. At its best, his writing has the same appealing style as that of Loren Eiseley, and the humanity of his motives is never in question.

Indeed, by the end of the book his deeper project becomes clear: namely that in the long run—in the *cosmic* long run, that is—the human species will have to reach out to the other intelligent beings that presumably occupy their own distant corners of the cosmos, and make itself part of a larger cooperative alliance of intelligent creatures. (Recall how Buffon, too, speculated about the living creatures which—he presumed—occupied the other planets and satellites of the solar system.) So Sagan's quest for ways of communicating with extraterrestrial beings ceases to appear a mere technological fantasy, and reveals itself for what it is: the pursuit of a "wider union" between humanity and its fellow inhabitants of this and other galaxies, to serve as the turning point by which human or terrestrial history can be finally embedded into, and integrated with, cosmic natural history.

Quite aside from all questions about the merits of Sagan's conclusions, to which I shall return, it is good to see this kind of book being written at all. Over the last fifteen years, most public discussion of science and scientific ideas has been narrowly restricted to questions about the technological "fruit" of scientific advances; and the enterprise of science itself has as a result been made a whipping boy by critics whose attention would have been better paid to the sins of industrial monopolies and the feebleness of America's system of government. So it is more than time for us to be reminded that the central concerns of much modern science have been cosmological, and that through most of its three-hundred-year history science has been associated far more closely with theology than with technology.

Natural philosophy has at all times been the product not of greed but of wonder. Its underlying appetite has been one not for more and cheaper consumer goods, but for a more comprehensive and comprehensible vision of Humanity and its place in Nature. (Thomas Henry Huxley's more euphonious phrase—Man's Place in Nature—has of course been overtaken by events.) In his attempt at a synthesis of all the sciences that can throw light on our origins and prehistoric evolution, from the first mammals of the Triassic period through the primates and early hominids to the human species as we know it, Sagan is accordingly writing in an ancient and distinguished tradition.

The "decentering" from local, human concerns that is required, if we are to follow Sagan into these speculations, will not come easily to many people. For much humanist thought about our affairs takes a basic position toward history and society that is, by Sagan's standards, entirely parochial. Its locus of attention is terrestrial; its historical span of attention is limited to a few decades—can we make it through the half century to the year 2027?—and it continues to set human life and affairs apart from those of other species and beings. Whether to practical people who are preoccupied with oil prices, population pressure, and the balance of power, or to humanist intellectuals—Leavis or Lévi-Strauss, Marshall Sahlins or Robert Heilbroner—the suggestion that we should put this secular, planetary, species-directed parochialism aside in favor of a more comprehensive astronomical/ historical perspective may seem frivolous and misdirected.

Yet our view of things will surely remain only partial unless we are prepared, and able, to view our affairs in that larger perspective also. It is, after all, not so long since "realists" were resisting the argument that human affairs need to be considered and analyzed as global or continental problems, instead of being dealt with only as concerning individual nations, city-states, or local communities. We should surely be ready to entertain all reasonable and scientifically justifiable thoughts about the place in nature, not merely of man or humanity, but also of the earth, of our sun, and ultimately of the entire galaxy.

This clash of perspectives between cosmologists and humanists is, of course, an old story, which goes back at least as far as classical Athens. Though ridiculed by Aristophanes for his supposedly excessive interest in astronomical matters, Socrates himself in fact abandoned the pre-Socratic philosophers' tradition of cosmological speculation, and refocused the philosophical debate on human problems and values. In very much the same spirit, Michel de Montaigne laid one of the foundation stones of modern humanism when he dismissed as presumptuous all attempts to make rational sense of astrophysics. Instead, Montaigne placed at the center of his position a motto which is often quoted as a sign of his liberality, but was in truth intended as a restriction of human inquiry—*Nihil humanum a me alienum puto*—"I deem nothing foreign to me, so long as it has to do with human affairs." And those of our contemporaries who have committed themselves to matters of human welfare and relevance, at all costs, will probably share the irritation that Sagan's continued devotion to historical cosmology would have provoked in Montaigne.

By contrast, those who have immersed themselves in the cosmological debate, as it has developed from Thales and Anaximander up to Hoyle, Wheeler, and Sagan, are able to lift their eyes from the human perplexities of the 1970s, to contemplate the larger world from different points of view, and to measure its affairs against very different scales both of time and of space. They not only "have the future in their bones"—to quote C. P. Snow—they have the remote past and the most distant parts of the universe at their fingertips also. Having a sense of the boundary conditions that our cosmic situation imposes on the development of human affairs may not, of course, contribute much to the making of policy, or the determination of short-term attitudes; and to that extent the wording of Snow's phrase is perhaps too portentous. Yet this, too, should be one element in the formation of a genuinely "humane" view of human affairs; and one important virtue of a book like Carl Sagan's is the contribution it makes toward reconciling the parochially human and the cosmological perspectives.

To say this is not to accept Sagan's argument entirely as it stands. In certain respects his background in the physical sciences betrays him, and he misses some important biological tricks. For instance, he has apparently fallen too completely for the molecular biologists' propaganda view of genetics as a mere sub-branch of biochemistry, in speaking of human beings as

> to a remarkable degree, the results of the interactions of an extremely complex array of molecules.

In spite of his insistence on keeping the genes and the brain in balance, he does not do full justice to the point he cites from Sherwood Washburn—that

> Much of what we think of as human evolved long after the use of tools. It is probably more correct to think of much of our structure as the result of culture than it is to think of men anatomically like ourselves slowly developing culture.

For what is true about our anatomical structure on the gross scale goes equally for our structure on the subcelluar level. In some respects, our "genes" are what they are as the outcome of evolutionary changes that involved crucial *cultural* elements.

The point goes beyond Sagan's argument to the whole debate about sociobiology and related topics. For instance, E. O. Wilson's explanation of the continuities between human social organization and the patterns of collaboration found in other social species links social theory to "genetics," but only in the sense of *population* genetics. And, as Wilson himself concedes, the relations between population genetics and biochemical genetics are speculative. Since evolutionary selection acts always on the "phenotype"— i.e., the fully developed, self-reproducing form—rather than directly on the "genotype"—e.g., the subcellular system of nucleic acids—the effects of cultural differences on the relative survival of different social populations (whether human or animal) may quickly become so great as to equal, or even in some cases swamp, the effects of biochemical differences. It is tool-using, language-using, culture-transmitting populations that are being preferentially selected; and the superiority of one set of tools, one repertory of language, one mode of culture over others is consequently capable of making a crucial difference to the survival of populations, even from the standpoint of genetics.

Wilson himself has sometimes talked as though a human being were simply "the gene's way of making another gene"—i.e., as though the survival of the "gene pool" were the controlling factor in all social evolution—and Sagan's argument is at times open to the same interpretation. I would note here, simply, that "gene pools" are not the only entities that exploit individuals in the interest of their own survival. Institutions and sociocultural forms do the same. The devout believer is the Church's way of ensuring the survival of the Church; the loyal citizen is the State's way of ensuring the survival of the State; the scientific apprentice is physics' way of ensuring the survival of physics; and so on. Once we reach the level of considering the genetics of social and cultural populations, such as those of the human species, we thus have to recognize that the "phenotype" on which evolution operates is itself made up of social and cultural, quite as much as of morphological, elements. So we are obliged neither to oppose culture and nature with Marshall Sahlins, nor to reduce culture to nature with Edward Wilson. Rather, we can now see that, even from a biological point of view, culture has the power to impose itself on nature *from within*.

There are a few other minor issues over which Sagan might be faulted. For example: he cites Chomsky's speculations about universal grammar in support of the idea that the human brain was evolutionally preadapted for language. Yet he should have noted that Chomsky himself has always been very scornful of all attempts to discuss the evolutionary history of language, or the physiological preconditions for its use. It is not in the theoretical linguistics of the MIT group but rather in the work of aphasiologists and clinical neurologists that he should have sought the evidence he wanted of the connections between the character of human language and the detailed anatomy of the brain.[5] Toward the end of the book, again, the excursions Sagan makes into such current issues as the ethics of abortion and the definition of death will strike some readers as thin, irrelevant, and even a little stale. (A tough-minded editor would have blue-penciled them.)

One may find it necessary to take issue with Carl Sagan over many details, but the over-all force of his argument will survive a great deal of correction of the details. It is a pleasure to have someone writing about scientific cosmology once again so elegantly, intelligently, and with such literary flair, after the comparative drought of the last twenty years.

5. See, for instance, the admirable papers of Norman Geschwind, collected in the volume, *Selected Papers on Language and the Brain* (D. Reidel Publishing Co., 1974), especially the 1964 essay on ''The Development of the Brain and the Evolution of Language.''

Arthur Koestler (III)

I

Twenty-five years ago, the well-known political novelist and journalist, Arthur Koestler—who had been a Berliner in the early thirties, but was now domiciled in London—caught his admirers off guard by publishing a new book called *The Sleepwalkers*. The most unexpected thing about this book was its *subject*. It was a fat, detailed, and heavily documented treatise on the history and psychology of scientific discovery during the Renaissance. Yet the book was written with all of Koestler's familiar flair and punch. Heaven knows, it was *readable*! Also, the author made no attempt to disguise his feelings about the protagonists in his story. From his viewpoint, Copernicus and Galileo turned out to be ambiguous or unsympathetic figures. As for Johann Kepler: at times Koestler's sympathy for him amounted (it seemed) to outright identification, so that some professional historians of science were heard to mutter sniffily that the hero of this latest autobiographical novel was Johann Koestler.

They had something to mutter about. For Arthur Koestler had in fact done an admirable job. He had read Ptolemy and Copernicus with some care, and Kepler with real love. True: he had an axe to grind, a thesis to argue, but it was a legitimate—even if, at points, an extreme—thesis, and he put a great deal of serious labour into presenting his reading of the Renaissance primary sources in a way that supported his particular interpretations. From the scholarly point of view, therefore, the book was not one that could be simply

Originally published as "Arthur Koestler's Theodicy" in *Encounter*, February 1979.

ignored; and, in due course, the central sections of the book were reprinted in textbook form, as *The Watershed*, and widely used in general education courses at American liberal arts colleges.

Yet the question remained: What had happened to Koestler? How had the highly articulate, anti-Stalinist author of those striking political novels, *Darkness at Noon* and *Thieves in the Night*—the autobiographer of *Scum of the Earth* and co-editor of a classic collection of essays in anti-Communist conversion literature, *The God that Failed*—become so distracted from politics that he was writing, not just about science, but about *the science of 350 years ago*? Here was a puzzle. A glance back at Koestler's earlier history apparently did something to resolve it. After all, his readers reminded themselves, he had been a "science writer" for the Ullstein press in Berlin back in the days before Hitler. But why Ptolemy and Galileo? Why not, say, Einstein and Heisenberg? What was it that had had the effect of snatching Koestler, not merely out of his earlier political vein, but out of his century?

The years passed, and that question has remained unanswered. In some ways, indeed, the mystery has only deepened. For, though Koestler continued to write from time to time about matters of current politics and controversy—capital punishment or vivisection, Britain's post-War afflictions or the history of the Jews—his main effort went into a series of books directed at seemingly unpredictable targets in and around the general area of "science". First, there came *The Act of Creation*, which put forward a syncretistic but suggestive account of wit, artistic creativity and scientific discovery, as three faces of a single underlying human "faculty" that he called *bisociation*—the capacity (one might say) to put Two and One together, and get *vingt et un*. Next, there came *The Ghost in the Machine*. This was a complex and technical discussion of Mind and Body, full of polemical attacks on shallow behavioristic accounts of human mentality, on fashionable philosophical analyses of "mental events" and the like, combined with the beginnings of a constructive theory about the evolutionary history and functional eccentricities of the human brain.

Then, there was a Shandy-like digression. Koestler's discomfort over Darwinian evolution theory led him into a flirtation with the twentieth-century supporters of Lamarckism, and a third "K" was at hand to displace Kepler in his affections: Paul Kammerer, who had in 1919 published a book attacking the belief in "historical coincidence" called *Das Gesetz der Serie*, and finally committed suicide after being accused of faking some experiments on the pigmentation of toads. (Faced with Koestler's resulting book, *The Case of the Midwife Toad*, even his more sympathetic readers found Kammerer a less convincing love object than Kepler had been.) Meanwhile, Koestler was editing and publishing a collection of scientific and philosophical essays, called *Beyond Reductionism*: these had originally been prepared for a symposium at Alpbach, in the Austrian Tyrol, where Koestler has his summer home. And now, finally, we have Koestler's own summary and digest of what he has been up to this last quarter-of-a-century—a 325-page condensation of his philosophical ideas and scientific speculations, with the title *Janus: a Summing Up*.

II

Janus repays reading, and reading with care, even by those who cannot read it with love. There has always been something odd, and hard to grasp, about these scientific books of Koestler's. All of them have had a certain elusive, confusing quality about them, carrying multiple messages on several different levels, and leaving the reader in the same squinting confusion as a double exposed photographic print. It has been as though, in each case, he had written two or three separate and distinct books, and then muddled the chapters together between a single pair of boards.

In *The Sleepwalkers*, for example, Koestler gave us a careful and critical analysis of the investigations and considerations that led Johann Kepler and the other astronomers of the Renaissance to put forward, and defend, the new views about the planetary system and its relations to the larger cosmos. But he accompanied this analysis with a strong attack on a certain kind of scientific rationalism, which had (in his view) led people over the last few decades to exaggerate the logicality of scientific method and scientific investigation. Far from science making progress by conforming to set, formal procedures (he argued) scientists have commonly stumbled their ways to new discoveries, guided—like sleepwalkers—by

some "eye" inaccessible to conscious criticism or logical calcula-
tion. Yet it was never clear who Koestler was getting at with these
objections. His attacks might have carried weight against a few
philosophers of science of the Viennese "logical empiricist"
school; but it seemed scarcely fair for him to treat these minor
opponents as spokesmen for an entire established scientific orthodoxy. After all, most real working scientists have little patience
with abstract pedestrian doctrines about "scientific method". (Nobody has spoken more passionately about the imaginative character
of scientific work than Albert Einstein.) So, Koestler's readers
were left in uncertainty about the rights and wrongs of his arguments. On the one hand, he had evidently collected a fascinating
mass of historical and biographical material about Copernicus,
Kepler and their associates. On the other hand, his polemic against
the over-intellectualizing of science apparently misfired. So, how
was one to read the book?

His subsequent books have been flawed by very similar ambiguities. Whether writing about humour or art, the shortcomings of
behavioristic psychology or the workings of the human brain,
Koestler can always be relied on to produce a rich and complex
body of well-digested factual material. (His experience as a science
writer in the early thirties has not gone for nothing.) Yet, behind
and between the expository chapters, the reader is always aware of
an undertow: a rhetorical animus directed against some supposed
purblind cabal of hardline embattled scientists, determined to
maintain the grip of an outdated mechanistic orthodoxy. The Public
Mind has to be rescued from this Dragon; and St. Arthur will
always be there, just in time, to cut loose the bonds and carry the
heroine off to safety. Still, here again, it is never entirely clear who
exactly composes this cabal, or how they supposedly succeed in
imposing their authority on those fresh generations of young,
hard-nosed, and highly counter-suggestible apprentices who enter
the natural sciences every year.

In this respect, *Janus* is a real help. Now, at last, we can separate
out the superimposed photographic images, and get them into focus
one at a time. For now, at last, Koestler gives us a systematic
account of the various different theses or "causes" (some will say,
"prejudices") that have provided the head of steam behind each of

his previous books from *The Sleepwalkers* on. Furthermore, he does so in a way that lets us see how these various theses fit together within a larger intellectual framework. And it turns out on closer scrutiny that our earlier confusions were well justified. The fine-textured fabric of scientific argument that he sets out for us in each of the books represents only one of the superimposed layers that make up his overall picture of Nature and Humanity, and a comparatively superficial one at that. For behind and below the arguments about behaviourism and evolution and statistics there lies a deeper, and an essentially philosophical, view of the world. If we are to get to the bottom of the Koestler Problem, therefore, we must penetrate through the surface tapestry of "science" and identify the larger scheme of concepts and principles underlying it.

My aim in the present essay is to take the arguments of *Janus* and its predecessors, and try to show what common underlying considerations have apparently led Koestler to adopt the attitudes he does toward each of these different sciences he criticizes. Once this is done (I believe) it will turn out that Koestler's break with his earlier political ways of thought has in fact been neither quite as sharp, nor quite as complete, as at first appeared.

III

All the same, we have to begin with the science. We can find our way through to Koestler's deeper philosophical positions only where this fabric shows signs of strain. The task is not easy, since Koestler comes near to swamping us with fascinating and diverting scientific discussions, so we must keep our heads. Even the most concise summary and discussion of the issues he raises is liable to appear laborious, and I can only do my best to keep my account fair and lucid.

Koestler's scientific arguments revolve around six chief theses: three negative, three positive. I list his negative theses first, not to suggest that his intellectual aims are merely destructive—Koestler is a master of polemical sarcasm, but his purposes in *Janus* go far beyond discrediting his opponents—rather, because it is easier to see what the meaning of his positive doctrines is, by contrasting them with the positions that he is determined to undercut and sweep away. As so often in philosophy, the destructive work is intended only as preparatory ground clearing: *démolir pour mieux bâtir*, to adapt one of Koestler's favourite French phrases. And the deeper thrust of his arguments becomes clearer, if we look and see what kinds of constructive possibilities he opens up for himself, by dismantling the particular rival positions that he singles out for his fiercest attacks.

The three positions on which Koestler concentrates his destructive fire are all of them, in one way or another, included among the "crumbling citadels of orthodoxy" which he denounces as superstitions from the "*Zeitgeist* of reductionist philosophy which prevailed during the first half of our century". (As he sees the matter, the scientific establishment has closed its ranks around these positions, and continues to defend them *à l'outrance*, ignoring the powerful arguments that should long ago have discredited them.) They are, respectively: (1) behaviourist psychology, (2) neo-Darwinist evolution theory, and (3) the belief in historical coincidences.

(1) To begin with the most obvious—one might almost say, the sitting—target: Koestler has a field day demolishing the more extreme claims of the psychological behaviourists. These claims are, for him, the most obvious (and obviously ridiculous) illustration of "reductionism" in the discussion of human affairs, especially the mental life of human beings. Both J. B. Watson and B. F. Skinner come in for some well-merited knockabout humour: especially, for their attempts to find a place in their theories for the creativity of the writer, the artist or the scientist. What Koestler objects to most in Watson, for instance, is his claim that artistic creativity consists in the capacity to pick out a successful solution to a technical problem *after* a period of "random manipulation"— that, in Koestler's words, "the solution is 'hit upon' *by chance* after many random attempts"; and his further suggestion that the artist's choice of a solution is "reinforced" by "admiration and commendation, both his own and others". On the contrary, Koestler argues: we must suppose that, even though often unconscious, all the "combinatorial activities underlying creativity" have been somehow *directed toward* the eventually successful outcome from the start.

Let me make two brief comments, at this point. First: on the technical level, Koestler does an injustice to Watson by his manner of paraphrasing him. Watson does not in fact claim that the solutions to problems in art or science are "hit upon by chance". It is (he claims) precisely the discriminating selectivity a creative person relies on in recognizing that a solution answers to the needs of his problem which displays his "creativeness"; and this "selectivity in recognition" is by no means a random, or chance matter. Right or wrong, what Skinner and Watson are arguing is that the successful outcome of creative work depends *entirely* on this discriminating selectivity. Given that, it is of minor importance whether the "ideas", or *possible* solutions, between which the artist has to choose originated in the first place from "random" combinatorial activities or from "directed" ones. Fertility of imagination is all. So the most effective way of promoting creativity is to find ways of *speeding up* the generation of new possibilities, so as to increase the repertory from which the artist has to choose, rather than to try and find ways of "preadapting", or "pre-editing", those possibilities in advance to the demands of any given external problem. This conclusion may not be true in point of fact; but it is not as ridiculous as Koestler's caricature implies.

Secondly: on a rhetorical level, Koestler makes things too easy for himself, by focussing his attention on such popular books as Skinner's *Beyond Freedom and Dignity*. Although most "radical behaviorists" do have something of the *simplificateur* about them, just because they are engaged in a legitimate task of scientific analysis, I know few who would defend the wild generalizations and extrapolations characteristic of Skinner's popular writings. As a result, Koestler merely ignores the intellectually more powerful work that has been done by such serious minded behaviorists as Joseph Brady and Israel Goldiamond, for whom the public dust-up over Skinner's *B F & D* is little more than an irrelevant joke. (At times, indeed, this tendency of Koestler's to confuse popularizations for serious science over-reaches itself. For instance, he refers to a jokingly exaggerated passage in Desmond Morris's *The Naked Ape* as "obviously meant to be taken in all seriousness", and cites it humourlessly as illustrating one of "two impressive strongholds of reductionist orthodoxy"!)

I hasten to add one disclaimer: I, too, deplore the present state of much academic psychology in the West, and I would applaud anybody who was able to show convincingly what has sidetracked so many psychologists into their present collections of trivial dead-ends. So, I am in no mood to put forward Skinner and Watson as masters of wisdom. It is just that Koestler's demolition job misses the central targets. Instead of looking seriously at the things that went wrong *inside* Western psychology around the turn of the century—E. B. Titchener's over-selective reading of Wilhelm Wundt, and the like—he picks on the one single feature of behaviourism that gives him offence, viz., its insistence on the efficacy of *success* as contrasted with *foresight*, and relies on this alone to discredit the whole enterprise. In the end, as we shall see, this fact tells us more about Koestler than it does about Skinner and Watson.

(2) Koestler's second chief target of criticism is

"the neo-Darwinian (or 'synthetic') theory which holds that evolution is the outcome of 'nothing but' chance mutations retained by natural selection."

About the problems of evolution theory Koestler has a great deal to say, and much of it is very interesting. True: many of the problems that he underlines, as damaging to the claims of contemporary neo-Darwinism, are problems that evolutionary biologists of Darwinist sympathies acknowledge and are actively working on, with reasonable hope of success. Still, like the admirable science writer he is, Koestler does a lot to make these problems intelligible to the ordinary reader.

For instance, he has some excellent things to say about the role of animal behaviour in shaping the course of evolutionary change. Animal species are not merely the passive victims or beneficiaries of their habitats: they themselves do much to shape those habitats, and to determine the "selection pressures" that are operative on them in any given situation. (Anatomical fleetness of foot will save the antelope from the lion, only if it is also psychologically quick to take fright, and flight.) Whether instinctive or exploratory, or both, such modes of behaviour have a central part to play in any adequate account of animal speciation, and Koestler is right to emphasize their importance. None of this, however, is in itself damaging to neo-Darwinism, any more than it was to Darwin's own ways of arguing: much of the original debate about "natural selection" in the 19th century, in fact, dealt with just such issues. It is only when these behavioural elements are quoted as proving the necessity for Lamarckian, as well as Mendelian mechanisms of inheritance, that Koestler puts himself squarely on the opposite side of the fence to the neo-Darwinists.

Why should he want to do this? On the basis of the biological evidence alone, this is never quite clear. Indeed, Koestler confesses that at one point he nearly sold the pass to the neo-Darwinians through inadvertence. Alister Hardy's ideas about the role of "progress by behavioural initiative" in determining the course of evolution were so appealing to him that he gave them more credence than he should have done:

> When I wrote *The Ghost in the Machine* I found this theory rather attractive, but on second thoughts it reveals a crucial flaw, in still relying—though to a lesser extent than the orthodox theory—on random mutations to achieve the fantastically complex changes in the nervous system that are needed in order to insert a new habit or skill into the organism's native equipment.

But he recovered his equilibrium, and here takes care to cover himself:

> From a methodological point of view it seems preferable to assume that the insect-hunting skill of Darwin's finch became impressed on its chromosomes by some unknown process *because it was useful*—that is, by Lamarckian inheritance—instead of invoking once more the Darwinian mantra.

Notice two things here, for future reference: the slightly odd use of the term "methodological" in the second quotation—as though the preference for Lamarckian processes, even unknown ones, were simply *procedural*—and the deliberate use, in the first, of the phrase "random mutations" instead of "mutations" by itself. Once again, it is the supposed *randomness* of mutations in the neo-Darwinian theory to which Koestler objects: he finds evolutionary change unintelligible (even, irrational) unless genetic mutations are by some means (even, by some means as yet unknown) *pre-adapted* to the future needs and conditions of life of the organisms that will carry them.

(3) This same theme, viz. Koestler's rejections of "chance" and "randomness" in the affairs of the natural world, is even more clearly apparent in his discussion of probability theory, which is the third of his targets. How can the physical order in the natural world be intelligible, if it is the product of quantum mechanical processes that have only a *statistically probable* outcome? At some points in Koestler's arguments, he seems to be attacking the whole enterprise of statistics itself:

> How do the dogs of New York know when to stop biting and when to make up the daily quota? How are the murderers in England and Wales made to stop at four victims per million? By what mysterious power is the roulette ball induced after a glut of "reds", to restore the balance in the long run?

(Koestler knows perfectly well, of course, that probability calculations are concerned with the question, What is most likely to happen where large numbers of independent events occur *in the absence of* clairvoyant dogs and "mysterious powers"? These questions are in fact rhetorical.) At other times, quantum mechanics seems to be Koestler's target: in particular, the "lack of determination" characteristic of individual events on the quantum scale. From Koestler's point of view, at any rate, one cannot rest content with a physical world picture built around so indeterministic a theory alone. At the very least, we must develop also a parallel, complementary account of the manner in which Nature counteracts randomness and creates "order" out of "disorder"—an account, that is, of

> the morphic, or formative, or syntropic tendency, Nature's striving to create order out of disorder, cosmos out of chaos, as ultimate and irreducible principles beyond mechanical causation.

IV

The three positive theses Koestler puts forward as his constructive contribution to science—or is it philosophy?—offer us a very different way of viewing the world of human behaviour, organic nature and historical change. His is a world ruled by order and forethought, hierarchy and degree: a world flawed only by the redeemable, or remediable, shortcomings of human nature. He constructs it around (1) the notion of "bisociation" referred to earlier; (2) a representation of the natural order as a complex, pyramidal "holarchy", composed of Janus-faced units that he calls "holons"; and (3) the hypothesis of a "paranoid streak" in human beings, which acts as the Worm in the Apple of human affairs but proves, on closer examination, to be the outcome of an "evolutionary mistake" in the development of the human brain, and corrigible by pharmaceutical means.

(1) Bisociation can be explained quickly. It is the faculty in human thinking that directs the creative mind, whether in the realm of science, or art, or of wit. As such, it is Koestler's answer to the behaviourists, serving to eliminate the objectionable element of "randomness" in the behaviourist account of human conduct and creativity. This is in fact a suggestive and potentially worthwhile concept, for which Koestler has already received some credit in the professional literature. True, it is not without problems and ambiguities of its own. If we take Koestler as meaning, simply, that creative thinkers do not generate their ideas with complete randomness, but confine their search for solutions to the sub-class of established "possibilities" appropriate to the *general* kinds of problems they face, his argument is hard to resist. But, if he wants us to believe also that our faculty for bisociation can be relied on to select in advance—if need be, unconsciously—ideas that are pre-adapted to the *particular* problem in question, this again implies an element of clairvoyance that many would find too high a price to pay. (As to this, my suspicion is that Koestler himself would rather *welcome* scientific grounds for taking clairvoyance seriously, so there is a real bone of contention here.)

(2) The terms "holon" and "holarchy", too, are capable of serving a useful purpose in science, with certain qualifications.

During the last seventy-five or a hundred years, the natural evolution of scientific thought has finally permitted scientists to move beyond the kind of analytical studies that Goethe, Schiller and Blake condemned as the monopoly of the Newtonian method; and they have been able, as a result, to begin reconstructing a picture of natural objects and systems as ''wholes'', made up of units or ''parts'' related together in a kind of superordination and subordination that is at once structural and functional.

This change is evident in physics, with the development of crystallography and the theory of solids; it can be seen in chemistry, with the decipherment of complex organic molecules and macromolecules; and it reappears in biology, with the study of physiological systems inaugurated by Claude Bernard, and the rise of effective theories of cell biology. (Despite Goethe, indeed, this project was already there *in petto* in the writings of Newton himself, with his general concept of Divinely ordained ''active forces'' capable of maintaining providential harmonies in the relationships between the ''particles'' of Nature.) Up to now, however, there has been no commonly agreed term for describing the features that are shared by, e.g., atoms—in their relations to electrons on the one hand, and to molecules on the other—by cells—in their relations to chromosomes and mitochondria on the one hand and to bodily organs on the other—and by human beings—in their relations to their own limbs and organs on the one hand, and to social institutions on the other. In Koestler's terminology, all three types of thing alike can be referred to as ''holons'': i.e., units that share the crucial property of being, in their various ways, wholes as well as parts. And the larger system of relationships in which ''holons'' of different orders of complexity are linked together (e.g., chromosomes, cells, bodily organs, and the entire human individual) he would have us refer to as a ''holarchy''.

If this is put forward simply as a terminological proposal, it has genuine merits. There are some real occasions on which it would be useful for scientists to mark the Janus-faced character of different natural systems (as being both wholes with their own parts, and also parts of larger wholes) by the use of some common, recognized term; and I myself know of no serious rival to ''holon''. All the same, this usage is not likely to sweep the scientific field. Even though scientists of many kinds have begun to move away from their earlier analytical preoccupation with material composition, or "parts", into a more systemic and constructive consideration of "wholes", the basic character of their enterprise still requires them to pay less attention to general likenesses than to specific differences. For scientific purposes, it is usually more important to be able to tell different kinds of cells (or institutions, or molecules) apart than it is to lump all these things together as sharing a common "holon-hood". So, I would be personally inclined to predict that these neologisms of Koestler's will make their appearance more often in public lectures, and other similar "popular" presentations of science, than they will in the professional scientific journals. And this is not something that Koestler will be entitled to take umbrage at: it is, in the classical sense of Greek philosophy, a matter of anagke or "necessity". With that single qualification, holons and holarchies deserve a respectful nod.

(3) One particular feature of the overall cosmic holarchy, however, has apparently ended up being less than fully functional. Somehow or other, human beings have ended up with brains maladapted to their actual place in the holarchic system. Not for the first time, Koestler argues, biological evolution has led to the appearance of a species one of whose crucial parts is not up to standard. When *homo sapiens* was ready to appear, the necessary neurological resources were not available among his evolutionary precursors, and the human brain came out as a botched job. To be precise: the neural pathways linking the cerebral neocortex to the deeper parts of the brain were not up to the needs of human existence. As a result, the human species was left with two counterproductive tendencies. It lacked the equipment necessary to ensure that Reason could maintain a general control over Emotion; and, more specifically, it developed a liability to be carried away by emotions of loyalty to any larger group or institution. These deficiencies, which Koestler speaks of as the "paranoid streak" in human nature, are the source of the fanaticism, bloodthirstiness, belligerency and other mayhem that have disfigured human history; and, in the nuclear age, they threaten the very survival of the species.

This hypothesis is, of course, a very complex one, with components of several different kinds; and it is hard to comment on it briefly. We may begin at the end, with the psychological components of the theory. It certainly seems to be the case that human beings turned out to be too biddable and compliant to authority for their own future good. (Koestler cites Milgram's famous experiments on the limits of obedience to good purpose.) So, it certainly appears that we need to give more thought than we have done hitherto to ways of encouraging the contrary elements in human nature, which are capable of expressing themselves in autonomous, independent and self-reliant conduct: the elements that enable any human individual to resist the impulses to conformity and ''stand up to group pressures''. But it is much less clear what, if anything, these human frailties have to do with *paranoia*. If we are going to give these tendencies a psychiatric label, we would do better to see them as probably springing from obsessional or depressive, rather than from paranoid sources. (Paranoid individuals are, in fact, driven by a suspicion of other people's motives that leads them to separate themselves from the group, rather than to cling to it with the kind of fanaticism Koestler is concerned about.)

This criticism is not a piece of pedantry. For one of the problems with Koestler's science is his tendency to prefer sloppy phrasemaking to exactitude. When he says ''paranoid streak'', he does not really mean the term ''paranoid'' to be taken seriously—it is just a *façon de parler*. And when he talks, likewise, about Human Reason not having a proper control over Emotion, and associates this diagnosis with faulty connections between the neocortex and the deeper brain, it is once again unclear just how seriously he means to be taken. After all, he must know that such grand abstract nouns as Reason and Emotion need to be treated with greater circumspection than he displays. It is not as though any human behaviour were *purely* the outcome of ''rationality'' or *purely* the effect of ''emotionality'' or ''affect''. To parody Immanuel Kant, we might say, ''Affect without Cognition is undirected, Cognition without Affect is powerless.'' (Even the creative intellectual activities of a mathematical genius have an ''affective'' component.) As for the neurological element in Koestler's theory: identifying the neocortex as the seat of rationality and goodness, and the deeper parts of the brain as the seat of evil, stereotypic emotionality alone, is far too simple. This attempt to localize ''psychological faculties'' in particular parts of the brain is just too crude to fit the facts.

The notion of an ''evolutionary mistake'' is even more perplexing. On the one hand, given all the additional resources of Lamarckian inheritance that Koestler wishes us to accept, the evolution of a badly botched brain would certainly argue a seriously misguided failure of foresight, or preadaptation, in the evolutionary changes that produced the genus *Homo*. (If Lamarckism were correct, could not such a ''blunder'' have been avoided?) From the neo-Darwinian standpoint, on the other hand, it is not hard to see that group loyalty and conformism would have been *highly adaptive* at an earlier, formative stage in human history: those of our anthropoid precursors who lacked these psychological characteristics would quickly have fallen victim to their more aggressive, tightly-bonded cousins. So in what sense was this a ''mistake''? Rather, it seems to be an earlier evolutionary *success*, whose consequences we now need to grow out of!

V

Before we move on to larger issues, two last things about Koestler's general way of handling scientific topics need to be remarked on. To begin with, the basic ambiguity in all his expositions—are his arguments really scientific or philosophical?—is apparent in the ways he treats his opponents and his supporters, respectively. He speaks of his supporters (We) as a loosely structured group of people with open minds and good will, who differ about many questions and have no wish to impose their ideas upon others. By contrast, he portrays his opponents (They) as an established scientific cabal, organized in defence of an entrenched, but outdated orthodoxy. As an outsider to both parties, I can report for a fact that the people he castigates as ''dogmatic neo-Darwinists'' are quite as much of a mixed bag as the members of his Alpbach Symposium, being sharply divided about many issues, and in no position to patch together any agreed system of dogmas (even if they wanted to), while many scientists of other persuasions perceive his own team of right-minded scientists as Koestler's Alpbach Mafia.

The dogmatism Koestler attributes to "the scientific establishment" is, in fact, a rhetorical fiction. So far as matters of theory and doctrine go, a science like biology is always full of diversity. (Anyone who could get general agreement about basic theoretical issues from James Watson and Ernst Mayr, Richard Lewontin *and* Ed Wilson, would have performed a miracle!) Correspondingly, the historical changes by which such a science advances involve kaleidoscopic processes of temporary association and dissociation, with transient alliances and oppositions continually forming and reforming. So, in the actual business of working science, there is no place for people to be, permanently, either *For* or *Against* any general doctrine such as "reductionism". There are merely periods, and fields of research, in which it is timely and fruitful to use (say) analytical methods rather than methods of composition, or *vice versa*. As a result, working scientists are generally prepared to be opportunistic, both in their professional alliances and in their philosophical commitments. To insist that analytical methods of doing science are either *always* or *never* right and proper—that is, to embrace either Reductionism or alternatively Holism (or Anti-Reductionism) as an entirely general posture—is to leave science for metaphysics.

This same ambiguity in the status of Koestler's arguments is evident, also, in his very different ways of criticizing scientific views that he objects to, and agrees with, respectively. In playing off scientific and philosophical issues against one another, in fact, he resorts to something of a double game. When he is on the attack, he casts his scientific opponents in the role of philosophers, and then ridicules the supposed "philosophical implications" of their positions. In putting forward his own views, on the other hand, he strikes a scientific pose, and asks only that his novel ideas prove their worth in the explanations of the future.

In criticizing the neo-Darwinists, for instance, he resorts (p. 170) to the time-honoured device of pointing out that the basic concepts and terms of neo-Darwinian theory are *interdefinable*, and claiming that they are on this account *tautological*, and therefore *empty*. Yet this "interdefinability" is, in fact, a feature of *all* general systems of scientific theory—the same thing is true, for instance, of the terms "force", "mass" and "momentum" in Newton's mechanics—and it does nothing whatever to undercut the *explanatory power* of such theories. Indeed, when the tables are turned and it suits his own purposes, Koestler is quite ready to defend his own theories against just the same criticism. ("One of the tests of a theory," he says on p. 67, "is that, once grasped, it appears self-evident.")

Still: it is time to leave these disagreements over the status and validity of Koestler's detailed scientific discussions. The really significant points raised by his enterprise go far beyond all scientific minutiae. So, we must now turn and ask, "What is the undercurrent of philosophical criticism that informs all of Koestler's objections—whether against behaviourist psychology, or neo-Darwinian biology, or quantum mechanics, or statistical interpretations of history?" As will be apparent by now, one word continually reappears in all of Koestler's scientific books, as the focus of his objections: namely, the word "random". It is any suggestion of randomness, or chance, or indeterminacy, that awakens his distaste and attracts his condemnation; and we may now enquire into the deeper, philosophical roots of that distaste.

VI

At this point, a brief digression into the history of earlier philosophy will help us to recognize the ancestry of Koestler's position. There is a well-known family of philosophical positions which reject *historical contingency* as incompatible with the "rationality" of things. In everyday life, we frequently brush aside puzzling or distressing historical occurrences as having "just turned out that way"; and this commonsensical view of History, as having an essentially contingent component, is reflected in any number of familiar idioms—"You never can tell", "There's no knowing", "History is full of surprises", and so on. (American usage has a convenient term for this component: viz., the word, *happenstance*.) Some philosophers, notably many of the "rationalists" of the late 17th and early 18th centuries, have been unable to reconcile themselves to this way of thinking. Whatever *did* happen, they have argued, *had to* happen just as it did; and, if we are unable to see the deeper reasons for any particular occurrence, that is simply a limitation of our understanding. If, for instance, we fully grasped the essential "inner nature" of Julius Caesar, we would be able (Leibniz claimed) to see that the historical unfolding of that essential nature *necessitated* his crossing of the Rubicon.

The rationalists' resistance to "happenstance" extended from their philosophy of history into their view of science. An empiricist like Isaac Newton might be quite ready to accept the structure and laws of the World of Nature as he found them, and might attribute the fact that they are thus and not otherwise to "God's mere Will". God was entirely free, in Newton's view, to decide just when, how, and in what form His Creation was to take place. But Leibniz could not rest content with the arbitrariness of this view: "What if God had, by some whim, chosen to create an *irrational* world? Surely, there must be some *sufficient reason* for God to choose the particular structure and laws He did?"

The symbolic issue around which this dispute turned was the stability of the planetary system. The new physical theory of the planetary mechanics put forward in Newton's *Principia* strongly suggested that the Sun and planets *in fact would* continue to move as they do for the indefinite future, without falling into disorder; but it could not prove that this *must happen*, and Newton did not absolutely rule out the possibility that slight deviations introduced by the mutual interactions of the planets might conceivably add up, instead of cancelling out, and so end by disrupting the whole system. "But never mind," he added, "God could always step in if necessary, and restore the original equilibrium by His own supernatural Action."

This final comment filled Leibniz with scorn. "It shows a very low view of the Creator," he wrote to Princess Caroline, "to imagine that He may have to step in from time to time, like an imperfect watchmaker, and set his Creation to rights." Quite the contrary: from Leibniz's point of view, it was evident that God *must* have created His World in such a form that it operated perfectly—it would have been *less than rational* for Him to have done otherwise. And, if Newton's system of natural philosophy failed to reveal the "sufficient reasons" that underlay the Laws of Motion and Gravitation and made the necessity of the Natural Order apparent to the rational mind, so much the worse for Newton. As Leibniz saw it, any acceptable system of physics had to take its place within a larger Theodicy: i.e., a system of ideas capable of demonstrating the rational *rightness, propriety and necessity* of the Divine Creation.

In physics, this attempt to explain away the contingencies of empirical Nature in terms of rational necessities in the Divine Mind did not survive Kant's criticism. If the universal Laws of Nature had any ''necessary'' basis (Kant replied) this must lie in the structures of *human* thought, and there was no occasion for us to second-guess the Creator. Yet, when the theory of organic evolution was put forward some sixty or seventy years later, the same objection was raised all over again. Charles Darwin, too, gave an account of human and animal evolution that left the appearance of our species as a matter of happenstance. He offered some interesting and illuminating suggestions about how the human species *had in fact* evolved; but he did nothing to prove that this *must have happened* as it did. If the history of the Earth had followed a different course, there might perfectly well have ended up by being no *Homo sapiens*—even, no animal life—at all. It all depended on what the physical conditions were, and what variant forms of life happened to appear, at each time and place in the evolutionary sequence.

This feature of Darwin's theory was violently attacked, by both theologians and philosophers, in terms that anticipate Koestler's own criticisms of contemporary neo-Darwinism. Anglican theologians saw it as eliminating Divine Wisdom and Providence from the origins of the human race; while secular rationalists saw it as denying the Rationality of Nature. (In a Darwinian world shaped by so-called ''natural selection'', George Bernard Shaw declared, ''Only fools and rascals could bear to live''.) So, from the mid-19th century on, mankind was confronted by the awesome thought that, if past History had been different, Nature too would have been different, and there might well have been no such thing as mankind; while, if future History goes as it well may, Nature too will change in drastic ways, and mankind will yet cease to be.[1] Once again, this

---

1. On the first page of *Janus*, Koestler remarks—rather oddly—that ''mankind as a whole has had to live with the prospect of its extinction as a *species*'' only since ''the first atomic bomb outshone the sun over Hiroshima.'' Yet that gloomy prospect was already a familiar one, given the evidence of the geological record

Darwinian conclusion provoked massive denial. And the central element of *contingency* in Darwinism has long delayed the understanding and acceptance of Darwinian biology in countries strongly influenced by philosophical rationalism. This has been especially true of Lamarck's own homeland, France. There, one may see Jacques Monod's recent best-selling book on evolutionary biology, *Le Hasard et la Nécessité*, as an attempt to make Darwin palatable to a Cartesian culture. (After all, Monod argues, there *is* a kind of "rational necessity" to Darwinism; only, it is one that operates on the ecological, rather than on the biomolecular level.)

There is, thus, nothing particularly idiosyncratic about Koestler's attack on neo-Darwinism. It is, in fact, all of a piece with a much more general rejection of contingency or happenstance in the scientific world-picture. Darwinism is unacceptable to him because it only tries to show us how evolution *in fact* happened: it does not attempt to prove that it *must have* progressed beyond, e.g., "the rabbit, the herring, or even the bacterium," still less that it *must have* produced the genus *Homo*. Above all, it is unacceptable to him, in the same way as behaviouristic psychology, because the basic processes of variation or mutation seem to him to rest on "blind", "random", "chance" or "undirected" phenomena.[2]

---

alone, some 10 or 15 years before the publication of Darwin's *Origin of Species*. In particular, it informs much of the pessimism of the time, apparent in, e.g., Alfred Tennyson's great poem, *In Memoriam*:

> Are God and Nature then at strife
> That Nature lends such evil dreams?
> So careful of the type she seems,
> So careless of the single life . . .
>
> "So careful of the type?" but no,
> From scarped cliff and quarried stone
> She cries, "A thousand types are gone:
> I care for nothing, all shall go."

2. Those who *insist on* the "randomness" of biological mutations—whether to assert it, like Monod, or to deny it, as Koestler does—are generally arguing for philosophical rather than for scientific positions. Among evolutionists, Monod may make an issue of the "chance" character of mutations when arguing a Cartesian case, in a work of *haute vulgarisation*; but more authoritative treatises, such as Ernst Mayr's *Animal Species and Evolution*, tend to employ more exact language. When

Like all the other sciences whose lack of "directedness" Koestler complains about, then, neo-Darwinism falls short in the last resort in the same way, and for the same kind of reasons, that Newton's physics did for Leibniz. All of them accept the "contingency" of natural events too easily, and so fail to demonstrate the necessary Wisdom and Order, Rationality and Providence of Nature. In a word: by the standards of philosophical rationalism they are *irrational*, and so unfit to take their place within the larger framework of a rationally intelligible Theodicy.

VII

Still, the question arises: Why should anybody place so high a priority on constructing a Theodicy? Why should anyone find the contingent character of the empirical world so offensive? Why, that is, should scientists be required not just to *explain* the phenomena of Nature, but also to *justify* them as being "rationally necessary"? Here, the comparison between Koestler and Leibniz can be taken a step further. Gottfried Wilhelm Leibniz was born in Leipzig at the close of the Thirty Years War. He grew up in a country devastated by the 17th-century conflicts between Protestants and Catholics, so he could see for himself all around him the destructive effects of wedding theological dogma and local patriotism. The only hope of preventing further recurrences of fratricidal conflict, fuelled by

---

used in a strictly biological context, indeed, the very term "random" has a quite obscure sense. Suppose we take as our paradigmatic example of a purely "random" process the unpredictable disintegration of a single radioactive atom: biological mutations certainly need not be assumed to be "random" in that sense. On the contrary, everything in the evolutionary process may perfectly well be *caused*. It is just that the causes that produce mutations, and so generate new evolutionary variants, do not seem preferentially "directed" toward the future good of the particular populations in which they occur. Rather, all the most powerful physiological mechanisms of genetic transmission apparently operate in ways that *shield* the genetic material from environmental influences—especially those that operate on individual organisms during their lifetimes—and this reduces the scope for Lamarckian inheritance of characteristics acquired during those lifetimes. This observation is not, as Koestler apparently supposes, *incompatible with* truly "creative" evolutionary change. Quite the reverse: such processes as Alister Hardy's "progress by behavioural initiative"—which Koestler himself came so near to falling for—are quite capable of producing, by Darwinian means, results indistinguishable at first sight from those of Lamarckian inheritance.

religious loyalties, was to find a way of "grounding" the religious beliefs of all Christian denominations—and, ultimately, even the Oriental religions—on a common rational basis.

Leibniz was thus the first systematic *oecumenist*. Confronted by the results of an ideological catastrophe, he saw a rational reconciliation between the supporters of all religions as the only way of forestalling the End of the World, as civilized Europe had known it. Despite their strenuously intellectual procedures, accordingly, the deeper purposes of Leibniz's metaphysical inquiries were highly practical, even political. To accept the seeming contingency of things, and to permit shallow appeals to "God's mere Will", would invite fresh theological dissension and destroy the possibility of achieving an agreed basis for reconciliation. So, underlying Leibniz's *Monadology* and *Theodicy*, there was not just an intellectual, but also an ideological programme. Contingency in the operations of Nature and History represented for him, not merely a *philosophical misunderstanding*, but also a *political threat*.

Arthur Koestler was born into a world almost as confused and threatening as that which Leibniz knew. Growing up in Budapest in the years around the First World War, Koestler saw with his own eyes the collapse of the traditional dynastic system of European monarchies, and its replacement by a patchwork of squabbling nation-states. (Though the ideological themes used to justify the World Wars of the 20th century have been more secular than those of the 17th century Wars of Religion, future historians may yet come to know the period from 1914 to 1945 as the Second Thirty Years War.) Whatever its defects, the Austro-Hungarian Empire under Franz Joseph had been a stable and reassuring "holarchy", and those who once know this stability have never been wholly free of nostalgia for "the old days". So, how could any serious young intellectual grow to manhood in the wreckage of the Habsburg Era and not dream of some alternative, but equally comprehensive, system of international order? How could he, that is, avoid asking himself whether the international order might not be put on some other, more rational basis, about which people of open minds and good will from all nations might agree? And, if he was a young man of scientific inclinations, how could he avoid looking into, and being initially drawn toward, the claims of "scientific socialism"?

For such a one, the failure of Marxian theory to translate into humane political praxis, as chronicled in *Darkness at Noon* and *The*

God that Failed, was no local or isolated deficiency. It was a failure, at one and the same time, on the intellectual, the political and the scientific levels. In the Stoic sense of the term, it was a breakdown of "cosmopolis"—a failure of human beings to harmonise with the *logos* of Nature. Whatever virtues the theories of Marx and Engels might have had as pointers towards the structure of an Ideal Republic, these were not proof against the corruptions of real-life human nature. The same passions that Koestler had, in youth, hoped to see turned to Good were, in actual historical fact, all too evidently directed toward Evil. The temptations of political power combined with the influence of nationalistic fervour to form a witches' brew, and neutralized the forces of Reason in the political realm. So, experience with the actual historical realization of "scientific socialism" in the Soviet Union demonstrated to Arthur Koestler, as to many others, that practical Politics belongs to the sphere of Sin, rather than to that of the Ideal. If human nature had not been at fault—if we humans had not been endowed with some basic psychopathological deficiency, some inherited fanaticism, some "paranoid streak"—should not the Ideal Republic of the Reason, as dreamed of by the "scientific socialists", have been realizable in fact? Surely, if it had not been for this psychic deficiency, the God need not have Failed, the Noon would not have been filled with Darkness, the Night might not have been quite so full of Thieves.

VIII

I should perhaps (by way of a postscript) apologize to Arthur Koestler for construing his ideas in theological terms: sin, theodicy, the *logos*, and so on. Despite all his estrangement from political communism, he still remains as much of a scientifically-minded secularist as he was in his earlier socialist days. So, although the structure and motivation of Koestler's arguments may be illuminated by placing them alongside those of Leibniz, it is probably nearer the mark, psychologically, to think of him as a latter-day Stoic. (I say a Stoic, rather than an Epicurean, advisedly. It is his behaviourist and neo-Darwinist opponents who are the Epicureans of the contemporary scene, justifying personal equanimity by appeal to the moral indifference of natural phenomena.

Koestler, by contrast, sees no hope for us unless we learn to harmonise the underlying *logos* of human nature with that of physical nature, into a unified "cosmopolis".)

It is especially worth bearing the continued secular character of Koestler's thinking in mind, if we turn to look—in conclusion—at the therapy he prescribes for human disorders. Faced by the "paranoia" that he diagnoses as the root of our political afflictions, he turns to psychopharmacology as a source of hope. He comes (that is) to bring us not Grace, but a Pill. If the ultimate cause of our inability to establish a rational order in the political and social realm lies in the inadequate pathways joining our neocortex to the underlying parts of our brains, that is something which can be taken care of by straightforward biochemical means. We must just arrange for the pharmaceutical companies to develop a highly selective medicament, capable of preventing the "emotional" limbic system from dominating the "rational" neocortex; and we shall then be able—at last—to usher in the Age of Reason. For reasons temporarily beyond our control, the opening of the show may have been delayed, but now—though seventy-five years late—we shall be in a position to go ahead with it. For now we shall have the means to counteract group loyalty, misplaced obedience and other irrationalities that have hitherto stood in the way of a rational world-order.

How should this pharmaceutical agent be administered? There are a dozen possible ways, Koestler tells us. If the effects of the agent are really so "cool", chances are that, in this pill-popping age, the younger generation will latch on to it without needing to be pressured. Otherwise, it may simply be added to the municipal water-supply, like fluoride. "But why bother with such picky objections?", he asks. "That's just a straightforward problem of practical administration, and we shall be able to solve it as we go along. . . ."

Unfortunately, this response will just not do. And the fact that Koestler does respond in this way shows that he has come through all his earlier political experiences with the innocence of his youthful socialism untouched. Far from being a matter of "practical administration" alone, the problem of governing the utilization of his supposed "anti-paranoid drug" simply drives us back into the world of practical politics. For let us suppose that such an agent were available, and in mass production; and even, that its effects were all that Koestler imagines. How could we then ensure that the *political institutions* of all human societies and nations would operate in the ways required to prevent it from doing more harm than good? If there is one safe prediction to be made about the resulting state of affairs, it is surely this: that the political leaderships of the superpowers would ensure a supply of fresh, drug-free mountain spring water for their own Pretorian Guards, while the P.L.O. and the I.R.A., likewise, would find their own ways of preventing the revolutionary fervour of their followers from being sapped by municipal drug-therapy—even if administered in the cause of "rationality".

The sad fact is, that psychopharmacology by itself holds out no more hope of eliminating fanaticism and conflict from the political sphere, and giving us the means of building a rational world order, than Marx's scientific socialism did. If anything, the general availability of Koestler's drug would only accelerate the coming of W. B. Yeats's new Behemoth. Failing the creation of new and powerful institutions of international co-operation, it would lead to a situation in which "the best" finally lost the last of their "conviction", while the "worst" retained all of their "passionate intensity".

For better or for worse, then, the fundamental problems of political reform remain what they have always been: namely, to devise new institutional arrangements that will enable human beings, just as they are—whether flawed by Sin, or by cerebral inadequacy—to resolve their conflicts of interest and ambition in juster and less bloody ways than has been the rule in the past. In dealing with this fundamental political task, nothing is more urgent than to tackle the painful, piece-by-piece business of institution-building, with all the patience, intelligence and commitment to justice that we can command. Somewhere along this political road, there may yet turn out to be a place for Koestler's Pill. But, until the institutional reforms required in the political sphere have been carried much further, would not the availability of such a psychopharmacological agent be quite as much of a Threat as a Promise?

Gregory Bateson

In the literature and movies of the American Frontier the scout is usually depicted as a roughly clad eccentric who leaves the safety of the settlement and reappears unpredictably, bringing a mixture of firsthand reports, rumors, and warnings about the wilderness ahead—together with a tantalizing collection of plant specimens, animal skins, and rock samples, not all of which are fool's gold. At first the settlers find the scout's help indispensable; but once their community begins to consolidate he becomes a figure of fun; and finally, after respectability has set in, he is a positive embarrassment. Yet their premature respectability is vulnerable. When the settlement is struck by drought, the scout's nature lore leads the settlers to hidden springs of underground water, but once the crisis is past, respectability reemerges, and the scout is ridden out to the town line.

Within the world of the American behavioral sciences, Gregory Bateson has always had the scout's ambiguous status. He himself has never been an orthodox academic, either in his position or in his activities. With grants from the National Institute of Mental Health and other agencies, he has done his research in a Veterans Administration hospital in California, at the Oceanographic Institute in Hawaii where he studied the behavior and communication of dolphins, and most recently as a benevolent presence on the University of California campus at Santa Cruz. The disciplinary respectabilities of the academic world have meant little to him. For more than forty years, he has been publishing books and papers on any subject to which he had something to contribute.

He has written with equal fluency about animal behavior and

Originally published as "The Charm of the Scout." Reprinted with permission from *The New York Review of Books*. Copyright © 1980, Nyrev, Inc.

anthropology, communication theory and evolution, paralinguistics and schizophrenia. His achievements have challenged the professional ambitions of academic behavioral scientists in this country to establish self-contained "disciplines" within the human sciences as autonomous and well defined as those in the physical and biological sciences. Again and again, just when the professionals began to get themselves nicely settled, Gregory Bateson reappeared in their midst, with arguments to demonstrate that their theoretical and methodological certainties were uncertain. No wonder many of them have found his work exasperating as well as admirable.

Born in 1904, Gregory Bateson comes from the aristocracy of British intellectual life that Francis Galton described in such books as *Hereditary Genius*. His father, William Bateson, was a major figure in the revival of Mendelian genetics after 1900, and the Batesons moved among the Huxleys, the Darwins, and the other luminaries of English (particularly, of Cambridge) natural science. Gregory's own imagination quickly drew him beyond the boundaries of biology into anthropology; yet he has preserved a first-rate understanding of the biological sciences, which play a significant part in his new book. On a field trip to New Guinea in 1936 he met Margaret Mead, and since then his life has been centered in the United States. (Mary Catherine Bateson, their daughter, is herself an anthropological linguist.) Meanwhile, his intellectual curiosity and fertility have led him to build up for himself a circle of friends, such as Erik Erikson, who form a kind of American counterpart to Galton's intellectual constellation in England.

Gregory Bateson's background also did much to shape the problems that have been at the center of his thought. He was born at a crucial moment in the scientific debate about Darwinism. During much of its first hundred years, the Darwinian theory drew its main scientific strength from its power to account for the anatomical and physiological forms of living things. From the start, the most convincing physical evidence of evolution took the form of fossils: notably, the sequence of fossil forms by which the discoveries of paleontology were shown to correspond with historical geology. So much so that many people came to think of Darwinism as concerned, above all, with explaining such things as the giraffe's legs, the hummingbird's beak, and the coloration of moths.

Yet from the start it was clear that there were two missing elements in the theory as it stood in 1859: the full case for Darwinism must include, also, a convincing theory of genetics and heredity, and an account of the significance of behavior in evolution—that is, an account of mental or psychological evolution. Darwin's own theory of "pangenes" as the bearer of hereditary features left him, in certain crucial respects, a Lamarckian in his explanations: and it was not until the rediscovery of Mendel's work that the material was at hand for making serious progress in genetics. Meanwhile, though much of the early debate about evolution focused on behavioral issues (e.g., the evolution of instincts) and though Darwin himself published a book on *The Expression of the Emotions in Man and Animals* whose full importance is only just being appreciated today, psychological evolution remained largely obscure.

Obscure it might be, but it was also crucial. In a thousand ways, the behavior of living things can make all the difference to their success or failure in the evolutionary selection process. The food that is effectively available to a species in any habitat depends largely on its feeding habits; the giraffe outruns its predators in the wild only because it is biologically equipped to take fright, as well as flight; and the propensities to build webs, dams, and honeycombs are clearly as relevant to the historical fate of spiders, beavers, and bees as the shapes of their legs, tails, and stings. So once the solid foundations for a modern science of genetics had been laid by William Bateson, the outstanding weakness in the Darwinian scheme lay in the realm of behavior. In the long run, the Darwinian "natural philosophy" would carry conviction only if its categories could be expanded to embrace the mental as well as the physical, the psychological as well as the physiological aspects of human and animal nature; showing, for instance, how intelligence, communication, and symbolic expression, quite as much as drives, reflexes, and instincts, can be understood as "advantageous" products of evolution, and explaining all these different mental functions in both healthy and pathological modes of operation.

That has been Gregory Bateson's central mission. In one way or another, all of his notable contributions to science have sprung from his habit of viewing the mental life and behavior of creatures as functional, adaptive activities that need to be intelligibly related to their evolutionary history and habitat. So, in one phase of his work, he concentrated on animal communication, particularly the "language" of the dolphins. In another, he studied the ways in which living creatures improve their adaptation by learning; he was one of the first to point out the special power that comes with the evolution of a capacity for "learning to learn"—what he called "deuterolearning."

Elsewhere, Bateson played a pioneering part in the development of paralinguistics and kinesics: i.e., the study of the behavioral adjuncts and contexts of the use of language—including all those different ways in which our use or understanding of words and sentences cues in with our ability to "read" postures and gestures, facial expressions and hand movements, tones and inflections, emphases and hesitations. He was involved in a well-known collaborative project at the Stanford Behavioral Sciences Center on "The Natural History of an Interview" in the mid-1950s, which led to an elaborate system resembling musical staves, to display all the complex signaling modalities involved in the simplest exchanges.

Bateson's attempt to bring the concepts of semantics and semiotics to bear on the interpretation of behavior paralleled the program of contemporary structuralism and avoided some of its theoretical rigidities. It has also served Bateson well in two other fields. In psychiatry he invented the "double bind" theory to explain how failures in family communication can provoke mental illness: if a parent's words and nonverbal messages are sufficiently inconsistent and contradictory, a child can be put into a "no win" situation from which the only available exit is into psychopathology. (This theory is now an accepted element in the conceptual repertory of much family and other psychiatric therapy.) In Bateson's anthropological work, again, he never remained content with labeling a customary dance or ceremony, initiation procedure or mode of dress as "functional," just because it fitted in with the overall pattern or "structure" of the culture. Like a good evolutionist, he has always demanded to know, also, *how* it was functional—what message it communicated, what skills it transmitted, or how else it contributed to the viability of the culture, regarded as a successful, well-adapted *Lebensform*. For Bateson, that is to say, it has always been essential that culture and nature should each make sense from the viewpoint of the other. But he is no reductionist: he is as quick to find "cultural" elements in nature as he is to point out "natural" elements in culture.

To list Gregory Bateson's achievements in this fragmentary way is, however, misleading. For it distracts our attention from the integrating themes of his work, and makes it appear scrappy. Yet how else can one convey these themes? There is no Bateson's Law or Bateson's Theory, no formula to represent his unique thought, as $E = mc^2$ does for Einstein. What links all Bateson's innovations together, and what his younger associates have been drawn to in his work, is not so much a comprehensive theory about the phenomena of mental evolution as a systematic approach to its problems. In all our thought about human questions—whether about psychology or social science, politics or mental illness, education or language—we should (he insists) never ignore the evolutionary aspects of the questions of who we are and where we are. We should never forget, that is, to ask how our modes of living and thinking, talking and acting contribute to our success or failure as members of populations of natural beings, or fail to consider in what ways other natural beings, too, may share in the same mental heritage of intelligence, communication, and social organization.

So, in the present *Mind and Nature: A Necessary Unity*, as in its predecessor, *Steps to an Ecology of Mind* (1972), Bateson attempts to redirect our ways of dealing with human problems in an evolutionary direction. Centuries of formal logic and metaphysics developed in the service of an essentially ahistorical cosmology have (in his view) set our ways of thinking and talking into fixed molds and patterns of kinds that make it hard for us to adopt such an evolutionary approach. In his first few essays, therefore, he seeks to discredit and dismantle the rigid forms of thought generated by that earlier alliance of Aristotelian logic and ontology, Biblical history and pre-Darwinian taxonomy: forms of thought that encourage us to assume that all the basic processes of nature lend themselves readily to unambiguous description, permanent classification, and scientific analysis according to a "linear" conception of causality.

In a chapter called "Every Schoolboy Knows," he attempts to formulate a view that is better "geared to . . . the biological world," offering a series of homely examples to show the fallacies of a scientific view based on coding, conventional description, and quantitative measure divorced from the object or phenomenon that is being measured. Thus every schoolboy knows, or should know, that science "never proves," but merely "probes"—and probes only as accurately as available instruments permit; like the microscope and telescope, all "improved devices of perception will disclose what was utterly unpredictable from the levels of perception that we could achieve before that discovery." Moreover, "there are large classes of phenomena where prediction and control are simply impossible":

> Under tension, a chain will break at its weakest link. That much is predictable. What is difficult is to identify the weakest link before it breaks. The generic we can know, but the specific eludes us.

Or, to take another typical example, in defiance of the apparently logical notion that "nothing comes of nothing," successful evolution can depend on the meaning of zero:

> The letter that you do not write, the apology you do not offer, the food that you do not put out for the cat—all these can be sufficient and effective messages because zero, *in context*, can be meaningful; and it is the recipient of the message who creates the context. This power to create *context* is the recipient's skill. . . . He or she must acquire that skill by learning or by lucky mutation, that is, by a successful raid on the random. The recipient must be, in some sense, ready for the appropriate discovery when it comes.

Bateson understands very well that commitment to a static ontology and taxonomy was the means by which Aristotle was able to harness his newly invented syllogistic process to his scientific world view, and so make formal logic the prime instrument of scientific explanation. It was no wonder that his philosophical successors fell into the habit of construing causal connections as though they were logical connections, so confusing the physically necessary and the logically entailed. Within an evolutionary world picture, by contrast, no descriptions can be trusted to hold good indefinitely, classification systems are all in shorter- or longer-term flux, and our deductive conclusions can be trusted only to the extent that natural events have in fact the character of "convergent" rather than "divergent" sequences.

Bateson's constructive counterprogram, as set out in the central essays of *Mind and Nature*, is built around three notions—the necessity for "multiple descriptions" of all natural processes, a "circular" conception of causal interconnections, and the role of "stochastic processes" such as natural selection in generating new modes of adaptation. All of these notions he expounds here with a kaleidoscopic procession of illustrations and allusions. (Many of these are indebted—ironically, as I shall argue later—to the ideas and arguments of Bertrand Russell: notably, Russell's causal theory of perception and his logical "theory of types.")

A properly evolutionary way of dealing with experience obliges us to recognize that no event or process has any single unambiguous description: we describe any event in different terms, and view it as an element in a different network of relations, depending on the standpoint from which—and the purposes for which—we are considering it. Nor shall we usually be able to distinguish the "causes" among phenomena from their "effects": within the organic world, especially in ecological and evolutionary processes, chains of objects and processes are commonly linked together in circles or spirals, so that each of them is implicated in the causal fate of all the others.

The best we can do in such a case is to understand all the interlinked chains within which our affairs are caught up, and consider how they might be modified so as to operate more advantageously as wholes: that is to say, in such a way that these entire systems become better adapted. During much of the twentieth century, from Durkheim and Parsons on, the central conceptions of social and behavioral science have been modeled on those of physiology: *system, structure, function*. The time has now come, in Bateson's view—and it is hard to disregard his argument—to look for those central conceptions rather in evolutionary theory: *variability, selective pressures, adaptedness*.

If that is so, Bateson argues, we must take more seriously the significance of "stochastic" processes: those dual processes, familiar in a whole range of fields from Darwinian theory to economics, in which randomly generated variations combine with external selection procedures to establish new patterns of adaptiveness and "optimization." In this respect, it is helpful to contrast Gregory Bateson's position with its direct opposite, as presented most recently in Arthur Koestler's *Janus*. Koestler finds the notion repugnant that creative innovations and worthwhile novelties could spring from a series of essentially "random" variations: he denounces this idea wherever it appears—in behaviorist psychology or quantum physics, in neo-Darwinism or the historians' acceptance of contingency. Great new achievements cannot simply appear out of the blue! New forms of value (Koestler insists) must surely have been provided for beforehand: either by conscious foresight, or by selective imagination, or by some form of neo-Lamarckian causality. Yet that is just what Bateson is denying. Evolutionary ways of thinking, he argues, accustom one to the idea that true originality simply *cannot* be the outcome of straightforward planning or simple causality alone. Truly novel achievements can be recognized for what they are only after they appear. We then see that they have "proved adaptive" in ways that had not been "provided for beforehand," either conceptually or causally. And the road to wisdom in the future must begin with the acceptance of that kind of unpredictability in historical events—with the encouragement of innovation, on the one hand, and the better understanding of "adaptation" and "adaptedness" on the other.

To point out that Gregory Bateson does not give us anything approaching a comprehensive theory of mental evolution is not really to criticize his work. For Darwin's failure to deal fully with the psychological aspects of evolution was no accidental lacuna. Quite apart from the lack of any direct evidence about the behavior of living creatures in earlier epochs, there were some more serious obstacles to any extension of evolutionary ideas into the realm of mind. Darwin's own teacher, Adam Sedgwick, was only the first of many who feared the consequences of bringing the mental and moral aspects of human nature within the scope of the new evolutionary theory; and even today attempts to move in this direction meet with stiff philosophical resistance. (Sociobiology is only one example.) Before we can reach the stage of developing specific theories of mental evolution, as a result, we need some kind of philosophical reorientation: setting aside the unscrutinized habits of mind that stand in the way of any such extension.

In that respect, the issues Gregory Bateson is concerned with are central to the development of twentieth-century scientific thought and method. The program that generated what we paradoxically call "modern" science, from the time of Descartes and Newton on, began with an act of abstraction whose consequences it has been hard to escape. What Descartes required us to do was not just to divide mind from matter: more importantly he set humanity aside from nature, and established criteria of "rational objectivity" for natural science that placed the scientist himself in the position of a pure spectator. The classic expression was Laplace's image of the ideal scientist as an omniscient calculator who, knowing the initial positions and velocities of all the atoms in the universe at the moment of its creation, would be able to predict, and give a running commentary on, the entire subsequent history of the universe—but only from outside it. Such a posture is open to us in practice, however, only when the "coupling" between the scientist and his objects of study goes only one way—when he can observe how those objects are behaving without influencing that behavior in the process.

The most significant novelty in twentieth-century science, generally, has been the fact that scientists have run up against the limits of that Cartesian methodology at a dozen different points. As Werner Heisenberg showed us, the required conditions do not fully hold even at the finest level of physical analysis: there, our acts of observation alter the states of the particles we observe. The emergence of psychology as a self-sufficient science (or family of sciences) has equally threatened the traditional claims of Cartesian detachment and objectivity. Most of all—and this is where Bateson's work comes in—the development of ecology has made it clear just how far, and in how many ways, human life—not least, the life and activities of scientists themselves—is lived within the world of nature that the scientist is seeking to understand. We can no longer view the world as Descartes and Laplace would have us do, as "rational onlookers," from outside. Our place is within the same world that we are studying, and whatever scientific understanding we achieve must be a kind of understanding that is available to participants within the processes of nature, i.e., from inside.

Some contemporary commentators have, accordingly, concluded that the age of so-called "modern science" is past, and that we are now moving into a period of "postmodern science." (This phrase was coined by Frederick Ferré.) The point from which any "post-modern" science must start is the need to reinsert humanity into nature. Seen from that standpoint, many of Bateson's own claims (e.g., that "biological evolution is a mental process") seem less startling. Once we set ourselves seriously to the task of rebuilding the scientific world picture in a way that accommodates human beings—including scientists—along with all the other inhabitants of the natural world, the need to reintegrate matter and mind follows immediately: indeed, the supposed distinction between "material" and "mental" processes ceases to be terribly useful or fundamental for science.

What makes Gregory Bateson's work so significant is the fact that he has acted as a prophet of "postmodern" science, in this sense. The shortcomings he has seen in traditional behaviorist psychology and learning theory, in shallow interpretations of biological evolution, formal linguistics, in mechanistic approaches to psychiatry, and so on, have all of them sprung from his basic insight into the weaknesses of the Cartesian methodology as a program for future science. And this same insight explains, also, why he sees the first step toward the necessary philosophical reorientation of the human sciences as calling for a new epistemology.

While the agenda for Bateson's new book, like the agenda for his whole scientific career, has great merits—not least, philosophical merits—its execution is, all the same, flawed and incomplete. Some of the flaws are in his style and manner. While he can write wisely and thoughtfully, too much of his present argument is shrill and scolding in tone. (In this, he shows some of the less admirable features of a prophet.) By now, the sheer novelty of his program has surely worn off. What we need from a "postmodern" natural philosopher today is not more exhortations to change our ways: rather, we need a careful and detailed examination of what the new methodology implies, both for the separate sciences affected by this transformation, and for the overall integration of the human sciences with the sciences of nature. In this respect, *Mind and Nature* falls short of its proper objectives. Indeed, at many points the book seems unsure of its intended audience. Given the crucial character of its central themes, one might wish that Bateson had argued his case on a consistently higher plane. Too many of the essays (often with titles like "Every Schoolboy Knows . . .") are aimed at elementary, not to say sophomoric readers; and the tags that he chooses to expound ("Sometimes Small is Beautiful," "Nothing Will Come of Nothing" and the like) come across as exaggerated or trivial.

Why does he write in this tone? It may reflect the comparative isolation in which Bateson has lived and worked. Somehow, his background seems to have reinforced his sense that he did not need to "prove himself": his true colleagues all along have been not his contemporaries but his great precursors down the ages. As a result, he has cared too little about other people's opinions of his work: he could leave a clearer mark, and do more good, by engaging his opponents more closely. As things stand, the present book will strike some of his colleagues as shallow and patronizing; and there is a danger that he will, once again, provoke impatience rather than admiration from those who could most usefully listen to what he says. For those of us who respect his approach to natural philosophy, and who find many of his ideas congenial and appealing, that is a matter for particular regret.

There are also some very real difficulties in the content of his present argument. Early in the book, he sketches out the main lines of an epistemology whose neo-idealist themes ("Science Never Proves Anything," "There Is No Objective Experience") will be familiar to readers of such books as Thomas Kuhn's *The Structure of Scientific Revolutions*, Peter Berger and Thomas Luckman's *The Social Construction of Reality*, and Paul Feyerabend's *Against Method*. He backs up this "antiobjectivist" argument with considerations of two kinds. Some of them rely on the causal interpretation of perception that was made popular in the 1910s and 1920s by Bertrand Russell's *Our Knowledge of the External World* and

The Analysis of Matter. For instance, Bateson writes:

> When somebody steps on my toe, what I experience is, not his stepping on my toe, but my *image* of his stepping on my toe reconstructed from neural reports reaching my brain somewhat after his foot has landed on mine. . . . To that extent, objects are my creation, and my experience of them is subjective, not objective. . . . Our civilization is deeply based on this illusion [of perceptual objectivity].

Oddly enough, this dichotomy of "internal images" as against "external events" is a piece of undigested Cartesian doctrine which Russell himself inherited from T. H. Huxley and J. S. Mill, and which goes back directly to the seventeenth-century epistemological debate. Most of those philosophers and scientists who take evolutionary theory seriously in thinking about perception— J. T. Lettvin and J. J. Gibson are two whose names come to mind—have long since rejected that dichotomy, along with the whole causal theory of perception, in favor of a functional and adaptive interpretation of entire perceptual systems and processes; and Bateson would have been truer to his own central insights if he had followed their example.

In many ways, indeed, Bertrand Russell is the last philosopher one would have expected Bateson to choose as an ally. (C. S. Peirce would have been a happier choice). Neither in his epistemology nor in his logic did Russell ever show much sensitivity toward the significance of evolutionary ways of thought. His logic may be different from Aristotle's, but in its own ways it is just as ahistorical: and certainly, in its original context, his "theory of types" had no relevance to the problems of multiple description within an evolutionary world picture. As an epistemologist, also, Russell never strove to carry the debate about sensation and perception beyond Darwin: rather, he was concerned to take it back to where it was before Kant. And, in any event, his causal analysis of perception is, surely, prime illustration of just that kind of "linear causal thinking" that Bateson's own argument justifiably attacks.

Elsewhere Bateson follows through his critique of "objectivism" to the point of concluding that epistemology is always and inevitably *personal*. The point of the probe is always in the heart of the explorer. What is *my* answer to the question of the nature of knowing?

Yet this conclusion is opposed not merely to a physicalist and formalist epistemology and method, such as the Cartesian ideal of "rational objectivity through detachment" that he is right to reject. It is opposed also to any kind of critical procedure for science—if epistemology is "inevitably *personal*," why not biology, too?—and it lands Bateson, at least in words, in the same kind of extreme romantic individualism as that of Paul Feyerabend. The difference is that, in Bateson's case, this conclusion is arrived at only through the exaggeration of a basically sound position. For what he is most concerned to emphasize, here as elsewhere, is the nonexistence of any uniquely correct scientific point of view or mode of description: natural events and processes always lend themselves to a variety and multiplicity of descriptions, depending on one's point of view.

But what differentiates one legitimate scientific point of view or mode of description from another is not anything personal: e.g., the fact that this is *my* point of view and that is *yours*. Rather, it is the fact that scientists are always free to approach any set of natural events and processes with a variety of legitimate purposes; and each of these alternative approaches generates, as a byproduct, its own distinct modes of description and styles of explanation, which—for methodological, not for personal reasons—are never in direct contradiction with one another.

Still, these criticisms do not affect the main course of Gregory Bateson's argument; and the sorts of adventurous forays into the intellectual wilderness that have been his personal specialty make some exaggeration and fragmentariness in the final product almost inevitable. Like the keen-eyed scout he is, he has discovered terrain that future scientists will be exploring and settling for decades ahead, and he has brought back for our contemplation some fascinating and intriguing specimens. It is not always clear exactly what we should make of them. Some of them, no doubt, may even turn out to be fool's gold. But his new book still gives us a tantalizing glimpse of what, in the new era of "postmodern science," an overall vision of humanity's place in nature will have to become.

PART THREE

THE FUTURE OF COSMOLOGY
Postmodern Science
and
Natural Religion

The essays in this part are based on material originally prepared for the Tate-Willson lectures given at Southern Methodist University, Dallas, in March 1979, and for the John Nuveen lectures given at the University of Chicago Divinity School in April 1979.

All Cohaerance Gone

Anyone who chooses to write (as I propose to do here) about "postmodern science and natural religion" must not be surprised if the first reaction of his readers is one of incomprehension. For many people today will find the very meaning of that phrase entirely opaque. "Given the present state of science and theology," they will ask, "how is the phrase *natural religion* to be understood at all?" Surely, the fates of natural theology and natural religion were settled long ago; so how can we discuss them constructively today? If we invite contemporary scientists to renew their contacts with "natural religion" or "natural theology," just what are we asking science to get together?

Until a hundred or a hundred and fifty years ago, these questions would scarcely have seemed so puzzling. Isaac Newton, that devout Protestant mathematician, did not hesitate to discuss the actions and attributes of God, the Creator and Sustainer of Nature, in the context of his own scientific writings. For instance, in the General Scholium that he added to the second edition of his classic *Principia Mathematica Philosophiae Naturalis*, he insisted that "to discourse of God does certainly belong to natural philosophy." (The subtle balance between inertia and gravitational attraction manifested in the stability of the planetary system was only one of many respects in which, as Newton saw it, the operations of nature testify to the rationality of nature's Creator.) And when Joseph Haydn echoed the Psalmist in his own oratorio, *The Creation*,

> The Heavens are telling the Glory of God:
> The Firmament itself shows forth His Handiwork,

what many of Haydn's listeners called to mind was in fact *Newton's* handiwork.

For some two hundred years, indeed, it remained quite customary for academics, preachers, and educated people generally to follow Newton's example. They did not hesitate to use discussions about zoological and botanical classification as occasions to marvel at *The Wisdom of God, as Manifested in the Works of His Creation*—the title of a famous book by John Ray; to present the functional efficacy of the hand or eye as evidence of the providential origins of the human frame; or even, though more hazardously, to cite the discoveries of geology as vindicating the accuracy of Moses as a historian of the Earth's origin and development. By the late eighteenth century (it is true) the onward march of scientific discovery was beginning to raise difficulties for literal-minded interpreters of the biblical history of nature. So, such edifying nineteenth-century series as the Bridgwater Lectures trod the narrow line between science and apologetics with some delicacy. Yet, right up until the publication of Charles Darwin's *Origin of Species*, those innocent appeals to the Argument from Design, which had been characteristic of Protestant scientists between 1650 and 1850, still kept their charm and their audience. As a source of Protestant inspiration, the Works of God revealed in the operations of nature remained, as they had been since the sixteenth century, coequal with the Words of God revealed in Holy Scripture. It is only in the last hundred and twenty years or so that natural theology and natural religion have gone seriously out of fashion.

Yet was this really necessary? Granted all the unquestioned reverses that biblical literalists suffered at the hands of science from the eighteenth century on, need this really have been taken as discrediting the entire enterprise of natural theology? Or should those failures have led theologians, rather, to reorient their enterprise, and to redefine its proper tasks and concerns in ways more congenial to the contemporary scientific enterprise and so less vulnerable to scientific criticism? In particular, what prospect is there nowadays of giving natural religion and natural theology both fresh spirit, and also a more satisfactory scientific content? That is the prime question to be addressed today in this area of thought.

It is hard to discuss this topic for long without lapsing into the vocabulary of the divorce lawyers. Clearly, the first step toward any "remarriage" of science with natural theology must be to understand better why their earlier union was "dissolved" in the first place. In what circumstances, and for what reasons, did they part company? What mutual understandings and misunderstandings, conflicts of authority and divisions of labor, fueled their disagreements and led to their separation? How far were the reasons for their divorce unalterable and fundamental; how far were they, rather, temporary and possibly transient; and how far do they still hold good for us, now, in the late twentieth century? Or have we now moved, without recognizing the fact, into a new situation from which the reasons, circumstances, and divisions of authority responsible for that earlier breach have quietly gone away? There are in fact (as we shall see) some indications that natural science and natural religion have already resumed an irregular cohabitation, without yet having chosen to publicize their reunion. If that is so, two final questions will then arise: namely, "How could this have happened in the first place?" and "On what terms could a formal reconciliation be negotiated?"

Shelley proclaimed the poets to be "the unacknowledged legislators of mankind"; and, certainly, we often find poets responding first, and with most precision, to basic changes in human modes of thought and feeling. So, we may take as our starting point a familiar passage from John Donne's poem "An Anatomie of the World," in which he points to the revolution in astronomy going on in his own lifetime as confirming "the frailty and the decay of this whole World." By way of preface, the very date of Donne's poem is worth recalling. One might be tempted to assume that it was a response to the completion of this intellectual revolution; yet that was not so. The poem in question appeared in the year 1611, when Galileo and Kepler were actively engaged in their most creative work, but some seventy-five years before Newton published his *Principia*. Just the previous year, in 1610, Galileo had issued a little pamphlet, *Sidereus Nuncius* (*The Starry Messenger*), which reported with illustrations his very first telescopic discoveries, of the four satellites of Jupiter, and the rest. The little book had been an immediate sensation, and within five years it was being read as far away as Peking.

Galileo's book quickly reached England, where in due course it made a vast impression on the mind of John Milton, so it would not have been surprising if John Donne came to hear of it. At any rate, here are the well-known lines:

> And new Philosophy calls all in doubt,
> The Element of fire is quite put out;
> The Sun is lost, and th'earth, and no mans wit
> Can well direct him where to looke for it.
> And freely men confesse that this world's spent,
> When in the Planets, and the Firmament
> They seeke so many new; they see that this
> Is crumbled out againe to his Atomies.
> 'Tis all in peeces, all cohaerance gone;
> All just supply, and all Relation: . . .

At this point, many readers stop, on the supposition that the "just supply" and the "Relation" that Donne is speaking of refer back to the astronomy and physics discussed in the eight preceding lines. But they would do better to press on. As the punctuation indicates, Donne's references to *supply* and *Relation* in fact refer, not back to the astronomy, but forward to the next four lines, in which the topic of discussion suddenly changes:

> 'Tis all in peeces, all cohaerance gone;
> All just supply, and all Relation:
> Prince, Subject, Father, Sonne, are things forgot,
> For every man alone thinkes he hath got
> To be a Phoenix, and that then can bee
> None of that kinde, of which he is, but hee.

What are Donne's poetic antennae picking up in this passage? He is responding, first and foremost, to the close connection that his contemporaries saw between two issues that, to most twentieth-century readers, will appear quite independent of one another. On the one hand, the "new philosophers"—that is Copernicus, Galileo, and the rest—were engaged in dismantling the traditional astrocosmological image of the world: the classical scheme that was familiar to readers of Dante, as much as of Aristotle, with Earth at the center, Water overlaying it, Air surrounding that, and "the Element of Fire" (the so-called Empyrean) forming the substance of the all-encompassing Heavens. On the other hand, the people of John Donne's generation had experienced a breakdown in their sense of "Relation"—that is, their sense of natural status and relatedness. Once the earth was displaced so that it became one of the minor planets of the sun, instead of occupying the center of the cosmos, albeit its corrupted center, people lost their former sense of "knowing where they were" in the overall scheme of things; and the very idea of natural status was called in doubt along with everything else. Maybe there was no reason why these astronomical changes should have been thought of as discrediting the concept of "Relation" itself: still, the traditional astrocosmology had for so long provided the accepted framework for thinking about natural status that its demolition at first left people in confusion—no longer "knowing where they were."

Three aspects of this seventeenth-century transition will be worth examining here: (1) the issues that were directly involved in Copernicus's and Galileo's challenge to the older world-picture; (2) the intellectual and professional consequences that the transition had for the newer world picture that was eventually developed by Galileo, Kepler, Newton, and their successors; and (3) the implications of this transition for natural theology and natural religion. With these issues in mind, I shall hint at the direction of the resulting argument. The traditional astrocosmology was not the only possible basis for a sense of "natural status and relatedness": on the contrary, the accepted system of cosmic order (or "Relation") became tightly linked with astronomy only as the result of a historical accident. So, calling this particular astronomical scheme in doubt need not, by itself, have meant calling in doubt the entire sense of cosmic relatedness. All that Copernicus and Galileo had succeeded in demonstrating was that the particular tidy unity characteristic of the traditional picture was specious, and provided an unsatisfactory basis for ideas about "natural status." That being so, one alternative course of action open to their contemporaries would have been, not to reject the idea of "natural status" outright, but rather to develop a sounder foundation for it. As an outcome of further, equally accidental historical developments, however, it has taken us a long time to do so; yet today, after some three hundred and fifty years, we are at last beginning on that task. One central question for us to consider about the whole historical development of modern science must therefore be, "Why has the task of finding a sounder scientific foundation for natural religion and natural theology been so long delayed?"

Let us start by noting four main themes from the traditional astrocosmology: that is, that set of doctrines about the heavens and the earth which is often referred to today as "the medieval world picture," but whose origins in fact go back far into antiquity. This picture of nature did not even originate in classical Greece. In their own thinking about physics, astronomy, and related subjects, the philosophers of Athens were strongly influenced by preexisting systems of thought. Indeed, it is clear by now, from the work of scholars such as Bidez, Cumont, and Eliade, that all the main features of this world picture long antedated the peak of classical Greek culture: it formed, in fact, a widespread and accepted part of accepted Middle Eastern ways of thinking from before 700 B.C. When Alexander the Great finally conquered Babylon, accordingly, Aristotle's nephew Callisthenes collected and sent back to Greece all the material he could lay his hands on both about the astronomical computations of Kidinnu and his colleagues in Babylon (the official calendar-makers, whom the prophet Isaiah called "the monthly prognosticators") and also about the Babylonians' general theories of nature: for instance, whatever was available from the encyclopedic *Babylonika* of Berossos the Chaldean.

The first theme of this traditional picture, and in many ways the most significant one, was the divine character of the heavens. The *matter* of the heavenly bodies was generally equated with the most lively, active, and evanescent of the material elements, namely fire; and this element was widely thought of as forming also the substance both of the gods and of the soul. For the Babylonians, indeed, the sun and the planets *were* gods. Keeping track of the heavenly bodies was thus a pious duty as well as a practical task; and so it remained, as Festugière has made clear, in the later monasteries of Hellenistic Alexandria. In this first respect, indeed, there is a genuine continuity linking Zoroaster and his predecessors in preclassical Iran and Mesopotamia right down to the Stoics of Hellenistic times: the idea that the human soul shares its material substance with the stars (not to mention with "rationality") is seriously discussed by Plato himself, in the *Timaeus*, at least as "a likely hypothesis." Lingering influences of this same equation have remained alive in popular Middle Eastern thinking ever since. "How dark it is tonight," says Leila, the highborn and educated Coptic heroine of Lawrence Durrell's novel *Mountolive*:

"I can only see one star. That means mist. Did you know that in Islam every man has his own star which appears when he is born and goes out when he dies? Perhaps that is your star, David Mountolive."

"Or yours?"

"It is too bright for mine. They pale, you know, as one gets older. Mine must be quite pale, past middle age by now. And when you leave us, it will become paler still."

(While a literary echo is not, in itself, scholarly evidence, this is the kind of Alexandrian detail for which Lawrence Durrell has a trustworthy ear.)

The second theme we may pick up from the classical cosmology is, by our own standards, a more strictly astronomical one. As Otto Neugebauer has demonstrated, the forecasting procedures developed by Kidinnu and the other calendrists of classical Babylon were technically in advance of any that the Athenians ever invented for themselves: in this respect, the computations employed today at the United States Naval Observatory, in preparing tables of astronomical ephemerides, represent refinements of procedures first worked out by Kidinnu and his Babylonian colleagues, rather than of anything devised independently by Eudoxus or Heraclides or Aristotle. The astronomical preoccupations of the classical Greek philosophers were directed toward very different ends. They inaugurated our tradition of theorizing about astrophysics: "How far (they asked) can one make sense of astronomical objects and phenomena, in terms of ideas and principles derived from a study of more familiar terrestrial things and processes?" Given this initial divergence of aims, the major documents of later Greek astronomy, from Aristotle and Hipparchus through to their culmination in Ptolemy's *Megiste Syntaxis*, were so many attempts at integrating the Babylonian and Hellenic traditions into a single system: wedding the computative achievements of the "monthly prognosticators" with more speculative Greek ideas about astrophysics. Just how intractable that task was is shown, not least, by the long hiatus that ensued from the time of Ptolemy to that of Copernicus.

Third: we should remind ourselves just what transformed the traditional mixture of scientific astronomy and astrotheology into a genuine *cosmology*. The unifying factor was the Greek conception of *cosmos* itself: that is, the conviction that the entire system of the world forms a single, integrated system united by universal principles, that all things in the world consequently share in a common "good order," in short, that the universe or *ouranos* is "well turned out." (Not for nothing, the word for "good order"—*cosmos*—is the origin of our own word *cosmetics*, also.) As a result, the natural philosophers of the Renaissance and seventeenth century took, at the outset, one highly significant step away from cosmology. In classical times, it was commonly assumed that Humanity and Nature participate in a single, common order. At the end of Plato's *Republic*, for instance, Socrates points to the rational structure of the astronomical order (*cosmos*) as encouraging our expectation that the human social order (*polis*) can be put on a similar rational basis: if that were done, human affairs and natural affairs would harmonize into the single unified order that the Stoics were later to call *cosmopolis*. By contrast, all the dualisms and dichotomies that have been characteristic of science and philosophy since the year 1600—separating Humanity from Nature, Mind from Matter, Rationality from Causality, and the rest—were foreign to the thought of antiquity, and became influential only during the modern period. So understood, the world view of Descartes and Newton no longer represents a genuine *cosmos*. Instead, it is split down the middle; and, as Pascal liked to insist when he poked fun at the menial tasks imposed on Descartes's *Dieu philosophique*, it took the continual effort and attention of a benevolent God to prevent the two halves of the new scientific world picture from falling apart and to keep them operating together in harmony.

Fourth and finally: over the centuries, the traditional astrocosmology acted as a focus for human thinking—speculative, symbolic, and practical—about a wide range of different topics. Every human craft prefers to function under the best auspices—"by divine appointment," so to speak. Given the powerful significance of astronomical knowledge for agriculture, navigation, and statecraft, accordingly, there was no evident point at which one could draw a rigid line, or set sharp limits to the potential influence of the heavens. There was no evident reason, for instance, why Ptolemy, the author of the astronomical *Megiste Syntaxis*, should not also write a treatise on astrology, namely the *Tetrabiblos*. In due course, as a result, a highly complex system of "correspondences" grew up, linking celestial and terrestrial things of many different kinds. Each metal from base lead to noble gold, each part of the human frame from anus to brain, each family of plants or animals (and so on, and so on) operated under the high patronage of corresponding celestial beings. So, an intricate cobweb of influences apparently joined every kind of thing in the terrestrial microcosm to some counterpart in the celestial macrocosm; and a whole system of intellectual mirrors, artistic symbols, and poetic images came into circulation which was to shape the idioms, the iconography, and the imagery of Western culture and civilization, long after their origins were forgotten.

To come to the main point: the traditional astrocosmology performed many and varied functions at the same time. (One of its charms, indeed, was the fact that a single system of ideas appeared capable of serving so many varied human purposes.) Some of these functions were practical ones: keeping track of the changes of the seasons, forecasting solar eclipses, and the like. Others were purely theoretical: providing an intellectual basis for explaining the makeup and workings of the natural world. Others were symbolic and expressive: that is, iconographic. Others, again, were more strictly religious: testifying to the relationships between things divine and things human. The balance of emphasis as between all these varied functions differed from one culture and context to another. In Babylonia, practical functions on the whole weighed more than theoretical; in Athens, it was the other way about. Yet in none of the cultures of classical antiquity do we find people single-mindedly preoccupied with any one function, to the absolute exclusion of the others. Thus, in their own ways and on their own terms, Kidinnu and his colleagues in Babylon were capable astronomical theorists; while the Greek philosophers never ignored the practical significance of astronomical and cosmological issues, any more than the poet Hesiod had done. The great strength of the traditional world picture thus was the fact that it could be understood at once as an astronomical, a technological, and a theological picture.

Still, when all is said and done, and however deeply we may be tempted to immerse ourselves in the rich fabric of traditional cosmological thinking, two things must be put on record. First, the entire doctrine of "correspondences" which began with the identification of the sun and planets as gods, and went on from there—the entire program of using the facts of astronomy as a basis for determining "natural status" or, to use John Donne's words, "just supply" and "Relation"—grew up through a sequence of historical accidents. In a striking and imaginative essay, Henri Poincaré invited us to ask ourselves just how differently scientific ways of thinking would have developed if the sky had been perpetually covered by clouds. (In that case, human beings would have had no acquaintance with celestial objects or "heavenly bodies," and Newtonian mechanics would have had to be formulated without benefit of any astronomical data, such as Kepler's account of the mathematical ratios displayed in the observed movements of the planets. Indeed, Poincaré inquired, how would we in that case have figured out, even, that the earth itself *rotates*?) Poincaré's thought deserves to be carried further. Lacking the spectacle of the heavens, it is not merely scientific thinking that would have developed differently. Cosmological thinking, too, would have had to develop without benefit of astronomical observations, and the ancients would have had to express their sense, and their measures, of "natural status" in quite other terms than they did.

All the same—and this is the second point—there are good reasons to suppose that, even in the absence of all astronomical experience, some alternative cosmology would have taken shape, in which "natural status" was defined in terms of quite other relationships. No doubt, the new philosophers of the sixteenth and seventeenth centuries were right to condemn the unity of the older astrocosmology ("correspondences" and all) as a specious unity; but that did not rule out all possibility of finding some alternative and sounder basis for seeing all things in the world—human, natural, and divine—as related together in an orderly way, that is to say, in a *cosmos*. For the time being, however, people simply found it too hard to sever the long-standing, though accidental, links between cosmology and astronomy, or to turn their attentions to the further question that arose as a result: namely, what alternative form that cosmic order might take, if the heavens were painted out of the picture.

The merit of looking at the older cosmological world picture in some detail lies in the background it provides to the new scientific picture that displaced it. For we can understand why the strands linking science and natural religion became so frayed, from the mid-eighteenth century on, only if we already understand what had been the true basis of those links in earlier times. The new and more "modern" natural science, which was already beginning to develop when John Donne wrote his "Anatomie of the World" and which continued to do so right up to the early twentieth century, departed from earlier "natural philosophy" in some obvious and radical respects. This was not just a matter of the *content* of our theories about Nature. Those changes of content—beginning with the reorganization of the planetary system around the sun, instead of the earth—are striking enough, but were probably the least of the changes involved. To think of the transition from the classical world picture to the world of modern science in terms of content alone—as the replacement of Ptolemaic astronomy by the Copernican world system, of Galen's astronomy by Harvey's, and of Aristotle's mechanics by Newton's—would be to miss its real point. We would do better, in fact, to pay close attention (1) to the procedures characteristic of the methodology of modern science, (2) to the psychological attitudes associated with this methodology, and (3) to the changes in the social organization of scientific work, especially since the year 1800. For these other changes have, arguably, affected the relations between natural science and natural theology more profoundly than any of the specific novelties in the content of modern science have done.

True, some of the changes of content within science have been more influential than others, just because they also have had a wider methodological significance. The new philosophers of the seventeenth century were skeptical about the whole traditional system of "correspondences" just because it was, in their eyes, too tidy, too comprehensive, and too systematic to be believable. The vast cobweb of cosmic interactions that grew up around the traditional world picture was a product of fertile imaginations, rather than of controlled study and observation. Since there was really no way of telling what might discredit the supposition that the metal iron "went together with" the planet Mars, there was no way of telling what might establish that supposition, either. If experience was to be any guide, we would therefore do better to start by dealing with questions small enough, and phenomena isolable enough, to be selected out and dealt with separately, in abstraction from the rest of the universe.

The theory of "change" can serve as a useful illustration of this point. Medieval scholars had followed Aristotle's example by attempting to formulate an entirely general theory of change, embracing such diverse processes as the speeding up of falling bodies, the fading away of a bell's chimes, and the physiological aging of the human body. By contrast, the new philosophers abandoned that project, and confined themselves to studying specific types of changes, in particular kinds of things, independently and one at a time. If there was anything of a wholly general kind to be said about "change" as such, that would presumably become apparent subsequently somewhere down the road. For the moment, the more urgent thing was to avoid begging the kinds of questions the medievals had taken for granted, for example, when they assumed that the same laws of "uniform difform change" must be equally relevant to all such varied phenomena. So, from the late sixteenth century on, instead of seeking for a single comprehensive doctrine of change, natural philosophers began investigating the laws of kinematics, of metallic vibration, of physiological senescence, and all the rest, separately and as parts of separate sciences.

This change in the content and questions of the sciences was associated with an important change in its methodology and procedures. From the early seventeenth century on, and increasingly so as the centuries passed, the tasks of scientific inquiry were progressively divided up between separate and distinct "disciplines." As became increasingly apparent, the questions available for scientific investigation fall into certain natural types, or families; there are questions of mechanics, of optics, of botany and zoology, of chemistry, of electricity, and so on: and, for purposes of scientific investigation, issues of these different kinds are best sorted out according to type, and dealt with accordingly. The first step in any scientific inquiry thus became—and still remains—to identify the phenomena presenting themselves for investigation as (say) "gravitational" rather than "magnetic" phenomena, or phenomena for electrophysiology rather than cell biology. Until this has been done, there may be no way of knowing what questions are worth asking about the phenomena concerned, still less where we may eventually accommodate them within the House of Science.

This opening gambit requires (that is to say) an initial *disciplinary abstraction*, by which the phenomena in question are pigeonholed as belonging to this discipline rather than that, and so as topics for study and explanation in these kinds of terms rather than those. Every independent scientific discipline is marked by its own specialized modes of abstraction; and the issues to be considered in each discipline are so defined that they can be investigated and discussed independently—in abstraction from—the issues belonging to other disciplines. It may subsequently turn out that there are significant interrelations between different disciplines, for instance, those connections between biochemistry and genetics which are explored in modern molecular biology; but they are a later, second-order matter. On the primary and fundamental level, disciplinary abstraction and specialization became, and still remain, the first rule of effective scientific analysis.

At this point, we can identify one crucial difference between modern science and its earlier cosmological predecessors. The traditional cosmology was never preoccupied with any one isolated aspect or function of the traditional system in any exclusive or single-minded way: so far as cosmology was concerned, indeed, intellectual abstraction and single-minded preoccupation were just not "the name of the game." It is not that the traditional cosmology was consciously thought of as an "interdisciplinary" enterprise. Quite the contrary: historically it was *pre*disciplinary, functionally it was *trans*disciplinary, while psychologically—for those to whom it was a live option—it was even *anti*disciplinary. (That is just what the "new philosophers" had against it.) From the time of the Renaissance on, by contrast, the chief intellectual instrument—and virtue—of scientific work was, precisely, its single-minded preoccupation with the specific, narrowly defined questions proper to particular scientific disciplines.

As a result of this first kind of abstraction, the broad and general questions about "cosmic interrelatedness" which were the focus of the earlier debates about nature have been superseded by other, more specialized, disciplinary questions. This change, in turn, has had important consequences, both psychological and sociological, for the conduct and attitudes of natural philosophers: that is, for "scientists," as they came to be known from the year 1840 on.

Sociologically speaking, the work of science has become increasingly specialized and professionalized, as the pool of active working scientists has split up into subgroups devoted to the specific intellectual concerns of each particular discipline and subdiscipline. As matters turned out, the productivity of scientific intellectuals, quite as much as the productivity of manufacturing industry, proved to be subject to Adam Smith's generalizations about the efficacy of "the division of labor." As a result, over the past hundred and fifty years—precisely, the period when relations between science and natural theology reached breaking point—there was what Max Weber would call a "bureaucratic rationalization" of the ways in which the tasks of science are structured and its work organized.

This bureaucratization has been accompanied by a corresponding professionalization of the ethics of scientific work. By the 1920s or 1930s, scientists had developed tightly controlled and highly specialized habits of work, along with a very definite code of professional conduct, of a kind that would have been instantly recognizable to F. H. Bradley: a code of "my station and its duties" in which each working scientist's *station* was defined by his professional standing within the work of his particular specialized discipline. During the last fifty years, indeed, only a handful of natural scientists have been either versatile or imaginative enough to make significant contributions to more than two or three subdisciplines at most: in the professionalized world of twentieth-century science, the Erwin Schrödingers, Linus Paulings and John Desmond Bernals have been rare birds. For the most part, individual scientists have concerned themselves professionally only with questions arising within a single discipline: in consequence, the questions that might have been asked across the whole spectrum of disciplines have rarely even been posed, much less answered.

Meanwhile, the disciplinary work of science has also encouraged a novel psychological attitude on the part of the scientist toward his objects of study. If it is the scientist's job to adopt "for the purposes of the investigation" the particular viewpoint of some specialized discipline, it has been natural for the scientist to do so also with a certain detachment—a certain psychological distance, even a lack of "cathexis." Too much emotional involvement with his subject matter will not do the investigator's scientific work any good: warm hearts rarely go with cool heads. So, the procedures of modern science have come to call for a second kind of abstraction as well. Just as the topics to be studied in different scientific disciplines are considered in abstraction from one another, so too the scientist as a person is encouraged to enter on his investigation in a coolly intellectual spirit, and to approach his problem in abstraction from all other, more personal interests and concerns. If we did otherwise, we would risk allowing the reliability of our scientific results to be clouded and biased by the subjectivity of our other, nonscientific preoccupations. So, disciplinary abstraction within the work of the sciences has brought in its train, also, a certain personal abstraction within the minds of working scientists.

This second kind of abstraction—namely, the intellectual detachment required in a "scientific" attitude, as the price of objectivity—is a topic to which we must return later, in the second of these essays. During the twentieth century (we shall then see) scientific developments on many fronts have challenged any assumption that scientists can consistently adopt a fully detached attitude, even for their own scientific purposes; and, as a result, we are now moving into a phase of scientific thinking that differs from the science of the eighteenth and nineteenth centuries quite as significantly as the "new philosophy" of the seventeenth century differed from the scholastic science of pre-Renaissance Europe. For the moment, however, let us stay for a moment longer with that earlier transition; and ask, "Whatever became of natural theology and natural religion, as a result of the changeover from medieval to post-Renaissance scientific thinking?"

Initially, as John Donne observed, the effects appeared cataclysmic. People had linked the theological aspects of the traditional cosmology so closely with its detailed astronomical structure, and with the associated system of "correspondences," that the rejection of those details at once brought the associated natural theology into question also. In retrospect, of course, we might argue that many Christian theologians (at least) had brought the resulting discredit on themselves. They incautiously chose to stake their religious teachings and reputations too completely on cosmological doctrines that were quite peripheral to matters of genuine theological concern, and so were unable to disown those doctrines when the need arose. (It is interesting to note that Judaism largely resisted this temptation.) Once the older cosmology had been undercut, it was therefore necessary for Christians to reconsider the whole connection between scientific and theological issues in the study of nature. So began that unhappy phase in the development of Protestant thought which historians of ideas associate most immediately with the phrase "natural theology."

Many of those writers who were most active in the new scientific movements of the seventeenth and eighteenth centuries were also devout Protestants, and they saw their work for science as contributing equally to "true religion." (For instance, Robert Boyle described himself, in a charming phrase, as a "Christian virtuoso.") In their view, God's Hand had written the Book of Nature as surely as it had the Book of Scripture; and you could "read God's Mind" in the one as surely as in the other. So, these Protestant scientists decided that careful scientific observation would put them in a position to renegotiate the traditional alliance between science and theology, and became convinced that the advance of scientific understanding was to be welcomed, as strengthening their favorite theological instrument, namely, the argument from design. According to their new understanding, science simply undertook the laborious but necessary work of bringing to light the precise details of God's design for nature, while it was theology's responsibility to teach and interpret the divine scheme in more general, pastoral terms.

This proved an unfortunate decision. Having burned their fingers once by embracing one excessively specific and detailed "natural philosophy" as orthodoxy, theologians might have learned their lesson and found some way of putting distance between themselves and science. But habit was too strong for them. In their enthusiastic desire to "read God's Mind in the Works of Nature," Newton and his associates set themselves to the work with a vengeance and quickly ran up a dozen lines of theological credit that they could better have done without. Scientific investigation soon brought to light natural causes and origins for one after another of those features of nature that the seventeenth-century enthusiasts had identified as specific marks of divine choice and supernatural wisdom; and, as a result, the new version of natural theology—based not on the idea of cosmic order, but rather on that of divine design—turned out to be a wasting asset. The joint intellectual enterprise that Isaac Newton, Robert Boyle, John Ray, and their colleagues had entered into so keenly was, after all, set up in a way that mortgaged the interests of theology to those of science. Any successful investigations only had the effect of increasing the equity holdings of science, while the holdings of theology were progressively reduced, including only those residuary assets not yet incorporated into the new scientific system.

Once the professional enterprise of modern science was fully tooled up and running efficiently, according to the best economic principles of Adam Smith, the theologians' position in the joint venture became even more marginal and uncomfortable. In this respect, matters did not really come to a head until well into the nineteenth century. Many of the leading Protestant scientists of the eighteenth century (the Reverend Stephen Hales of Teddington, the Reverend Joseph Priestley of Birmingham, and the rest) were themselves in holy orders; so, quite aside from the amateur theologizing of the Newtons and the Boyles, the religious interpretation of the new scientific discoveries was taken care of as things went along. But, with the increasing professional fragmentation of scientific work which began around 1820 or 1830, the pressure was on. All the questions that arose within particular scientific disciplines had corresponding groups of people whose professional task it was to deal with them. By contrast, the integration of scientific results into an overall, transdisciplinary cosmology, and the theological interpretation of the resulting structure, were nobody's disciplinary concern; and, in an age marked by bureaucratic rationalization within all spheres of human activity, this soon came to mean that they were nobody's professional task, either. From 1860 on, accordingly, fresh generations of scientists arose who no longer knew or cared that the so-called laws of nature had originally taken their name from being "the laws decreed by God for His Creation," and who felt in no way bound by the implicit terms of the alliance between science and theology embodied in Newton's "natural philosophy."

Worst of all, from the mid-nineteenth century on, the principles and concepts in terms of which the intellectual concerns of the different scientific disciplines were framed came to be conceived of in ways that positively excluded those transdisciplinary issues that might, in other circumstances, have led on to a fruitful discussion of natural theology. In its actual content (that is to say) the science of the nineteenth and early twentieth centuries became an aggregate, rather than an integration, of results from its component disciplines. Anyone wishing to enter into an intelligent and intelligible discussion of mechanics, for instance, had to begin by mastering and accepting the relevance of such mechanical concepts as "inertia" and "momentum." Those few who were tempted to ask critical questions about the credentials of those concepts soon discovered that physicists professionally concerned with mechanics had nothing to say to them: "Could we do without *inertia*? What kind of a question is that?" To transfer a term of my own from a very different context, questions of natural philosophy, theology, and religion became from the standpoint of any single specialized scientific discipline, "limiting questions." Given the climate of the time, this meant that they also became "limiting questions" from the standpoint of *science as a whole*.

For anyone intellectually committed to a particular scientific discipline, the answers to such questions "went without saying." They represented, at most, the "constitutive tautologies" of that entire disciplinary activity, or scientific "form of life"—questions that could apparently be pressed only from intellectual confusion, or else in a spirit of machine breaking. The professional exclusion of theological issues from the day-to-day business of the scientific disciplines led, inescapably, to the conceptual exclusion of theological ideas from the realm of "scientific meaning" as well. So, the final outcome of the Newtonian Concordat was to drive questions of natural philosophy, theology, and religion out to the very boundaries of rational intelligibility—out to those *Grenze der Vernunft* at which Immanuel Kant had been inclined to locate them all along.

The changing relations between science and theology during the nineteenth century bring a vivid image to mind. The professional world of the industrial corporation being what it is in the late twentieth century, anyone in middle management is subject to a recurrent nightmare. You go away on vacation, leaving behind a flourishing and active department with clearly defined terms of reference, an established staff, a suite of offices, even perhaps—as the carpet advertisements put it—"a Bigelow on the floor." Having come back from the Bahamas two weeks later (the nightmare goes) you return to the office and walk in the door, to be met by blank-faced strangers. Your job, your department, and your office no longer exist. Nobody recognizes that you have any business there; there is no sign that your department is still functioning; and the very partitions in the working space have been shuffled around so drastically that you can no longer identify your own former suite of offices. You cannot even say, triumphantly, "Look! This is the *room* I used to work in," for there *is* no longer such a room to be recognized. (It is a film theme worthy of the Italian director Bertolucci.)

This nightmare is the ultimate nemesis of humanity under the reign of the division of labor; and what happened to natural theology in the years between 1830 and the early twentieth century came very close to it. Within a situation in which intellectual efficiency had decreed the subdivision of science into independent disciplines, the instrumental rationality of Max Weber's bureaucracy simply wiped out the former responsibility that natural theologians had accepted to the common enterprise on which they and the scientists had earlier collaborated. "You used to run a Department of Coordination and Integration, did you? Well, as you can see, we don't have any such department: all our enterprises run perfectly well, without needing to be coordinated and integrated. And now, if you don't mind, would you please go away, and leave us to get on with our work?" In short, the disciplinary fragmentation of science during the nineteenth century seemingly made the integrative functions of natural theology quite unnecessary.

Yet was that fragmentation itself final? That is the next question. The idea that science could operate forever as a simple aggregate of independent disciplines was never established absolutely or unconditionally. Arguably, it rested all along on certain presuppositions; and during the twentieth century these have turned out to have only a limited and conditional validity. Only from within a narrowly theoretical view of the scientific enterprise was its fragmentation into disciplines, or its subsequent reconstruction as a mere aggregate, acceptable at all; and, since the separation of natural theology from natural science—like the boundary between "substantive questions" and "limiting questions"—was itself determined in a manner relative to the limits of the existing scientific disciplines, the whole program was open to question. Organizational, methodological, and conceptual barriers that were originally erected as the outcome of historical changes may accordingly have to be demolished again, as a result of further historical changes; and, once we find reasons for questioning the orthodox "disciplinary" structure of science, it may yet prove possible to transform the neglected issues of natural theology from being limiting to being substantive issues once again. In some respects, furthermore, this has already begun to happen; and to that extent fresh opportunities are already opening up to cosmology and natural religion, if only we are ready to follow them.

Death of the Spectator

The immediate reason why natural science and natural theology parted company in the nineteenth century was, thus, the fact that by 1860 the natural sciences had become fragmented into a number of largely independent disciplines, each with its own specialized questions, preoccupations, and methods of inquiry. As a result, it was no longer the professional business of anybody in the sciences to think about "the Whole"—that is, to deal with those broader questions, about the overall interrelations between things of vastly different kinds, which had been a major concern of earlier cosmologies. Behind that immediate source of separation, however, there lay another, deeper reason for the breach between science and natural religion, to which we should now turn. It was possible for cosmological issues to drop out of sight, and out of science, in this way, only because the intellectual attitudes of natural scientists had become predominantly "theoretical"—that is, detached and self-consciously "objective"—so that nobody in the sciences any longer needed to think about "the Whole." This claim calls for some explanation.

Let us step back. In several of his plays, that fine contemporary dramatist Tom Stoppard has developed variations on a theme that may be captured in the phrase "Death of the Spectator." In *Rosencrantz and Guildenstern Are Dead*, two young men who appear in Shakespeare's *Hamlet* only as peripheral characters are thrust into the center of the action: from being onlookers at the skulduggery in the royal court of Denmark, they are unwittingly transformed first into participants in, and finally into victims of that skulduggery. In *The Real Inspector Hound*, two drama critics are watching what appears in act 1 to be a drawing room farce; but, by act 3, they find themselves sucked into the action of that same play, which now becomes a tragedy rather than a farce. In *Professional Foul*, an English academic philosopher goes to a congress in Eastern Europe intending to deliver a highly "objective" paper in analytical ethics; but he ends by throwing away his prepared script and smuggling out of the country a subversive manuscript written by a former graduate student in trouble with the political authorities. Stoppard's command of the English language reminds one of Joseph Conrad—Stoppard is himself a Czech refugee, and British only by naturalization—and, in all these plays, he shows us would-be spectators, would-be onlookers being transformed, half against their wills, into participants or agents in the very activities that they had initially been only observing. Any one of the plays could, thus, take as its motto Heidegger's maxim that, from now on, we can no longer regard the World simply as a View. The option of being mere spectators is no longer open to us: we are all of us, willy-nilly, agents in all that we observe.

Stoppard's work provides a good point of departure for us here because, like John Donne three hundred and fifty years earlier, he is in this respect picking up a fundamental shift in the character of intellectual life and attitudes. From A.D. 1600 onwards for some three hundred years, the central leitmotif of much self-consciously "progressive" science and philosophy was the need to pursue "rational objectivity" of a kind that could be arrived at only by a detached and reflective observer. Cartesian dualism made canonical a split in our vision of the world, which had the effect of setting rational, thinking humanity over against causal, unthinking nature, and so enthroned the human intellect within a separate world of "mental substance." Given this initial standpoint, the human mind had the task of observing (and syllogizing about) the world of material objects and mechanical processes, but always did so *from outside it*. There was, supposedly, no more than the absolute minimum of interaction in the cerebral cortex, or in the pineal gland, between the human observer and nature observed. From the time of Descartes right up to our own century, as a result, the fundamental intellectual attitude embodied in science and philosophy was strictly "theoretical." The philosophers and scientists of post-Renaissance Europe, it seemed, were at last able to live the life of pure reflection (or *theoria*) which Aristotle had acclaimed as the highest good. They were, in a word, true *theoroi*.

We may pause here, at the outset, and look briefly at the etymology of the word "theory" and its cognates. In classical Greece, the word *theoros* was originally used for an official delegate who was dispatched from a city-state to consult the Oracle about some problem of city policy. Such a delegate was one who "had a care for the gods": that is, he was a "divi-cure," a *the-oros*. (The initial *the-* in *the-oros* is, thus, the same as the *the-* of *theos*, "theology," and other divinity words.) By extension, the same word came in due course to be used also for an official delegate dispatched from a city-state for other purposes, for example, to attend the Olympic or other intercity athletic Games: not to take part in them as a participant, but simply as an official overseer, observer, and symbolic representative of the city. ("I represented Megara at the Isthmian Games." "Did you run in the foot races?" "No, I was there as *theoros*.") Eventually, the word was generalized further, and used to refer to any spectator at the Games, whether official or unofficial, as contrasted with a participant: correspondingly, the abstract noun *theoria* began by denoting the activity of spectating, onlooking, or observing any activity or process, by contrast with intervening, participating, or being an agent in it. As the final step, the word achieved its familiar Aristotelian status: *theoria* came to refer to the detached intellectual posture, activity, and product associated with the philosopher's study, observation, and reflection about the world, by contrast with the *praxis* of the carpenter, the farmer, or the fisherman. So, from very early on, philosophy—qua "theory"—became essentially the reflective thought of a spectator; though, in view of the high origins and affiliation of the term, the philosopher was thought of as a "spectator" with a touch of class or official status—even with a touch of holiness about him. (While we are playing with philosophical etymologies, notice that the Latin counterpart of the Greek word *theoria* is *contemplatio*. Given their more down-to-earth conceptions, the Romans used as their root metaphor for philosophy, not that of a pure spectator but rather that of a surveyor, whose task is to mark out the area for a *templum*—an area for the subsequent performance of auguries by others. Still, even a surveyor remains a kind of onlooker, once the real action begins. The *contemplator* marks out the site for religious ceremonies: it is the *augur* who actually performs the divinations.)

With this conception of "theory" as our background, let us now consider the following three-stage argument:

1. At the time of the so-called scientific revolution in the seventeenth century, this conception of *philosophy as theory* was taken over from metaphysics into natural philosophy by the new "mathematical and experimental philosophers" whose successors we know today as "scientists." That much is not hard to demonstrate. If there had been any doubt about the matter, Alexandre Koyré's *Newtonian Studies* has shown us how close Newton's basic philosophical program for science stuck to the metaphysical position of Descartes, despite all their disagreements about the void, about vortices, gravitation, and other specific details; and what had been good enough for Newton was, from 1720 on, good enough for most of his successors. Much of the striking success of the modern scientific movement in the subsequent two hundred and fifty years can, in fact, be attributed to the power of this conception: the "theoretical" program and methodology framed for natural philosophy by Descartes and Newton in the event proved extremely fruitful.

2. During the last fifty or one hundred years, however, the success of this program has had one unforeseen outcome. We have finally come to understand both that natural scientists can no longer rely on this methodology universally—if pushed too far, it runs up against natural limits and restrictions—and also what the nature of those inescapable restrictions is. One of the best signs that our understanding of meteorology is truly "scientific" lies in the fact that we now understand why, in practice, it is frequently *not* possible to forecast the weather: so too, more generally, one of the most important points in favor of the Newtonian methodology is the fact that, now at last, we understand at what points, and for what reasons, its intellectual goals and ideals *cease* to be realizable.

3. In this way, the familiar conception of "pure" or "theoretical" natural science has finally run up against its own intrinsic boundaries. As a result, we are entering a phase in the development of science in which it is possible to press our understanding forward still further only at the price of modifying our accustomed scientific methodology and ideals; and one consequence of this change will be that the traditional questions of cosmology, about the "overall

interrelatedness of things,'' can once again be opened up for effective discussion.

For the moment, we may spend a little time exploring the ''theoretical'' character of the seventeenth-century scientific program, and the kinds of knowledge or understanding at which it accordingly aimed. So far, I have spoken of this attitude as something that scientists demanded of themselves in carrying out this new program: if one were to obtain ''true'' scientific results, it was necessary to view the world in an objective, detached, and universalizable manner, rather than from any subjective, personal, or particular standpoint of one's own choosing. For this reason, the men who founded the Royal Society of London in the 1660s deliberately eschewed politics and religion. They aimed at establishing the natural sciences on a nondenominational, nonparty basis, and were determined to avoid doctrinal disputations between (say) Royalist and Cromwellian, Catholic and Protestant views about nature. Their goal was, not a Christian crystallography or a Muslim mechanics but a natural science that could be understood and agreed on by people of all faiths and politics.

The objectivity at which the new natural philosophers were aiming, however, also carried with it implications about the topics that were suitable for ''scientific'' investigation. To come straight to the point: the kinds of objective knowledge and understanding aimed at in the new philosophy were, prototypically, the kinds of knowledge and understanding appropriate to ''objects''—notably, physical objects. To spell this point out, let us begin by contrasting two different kinds of knowledge. There is one kind of knowledge that we typically arrive at on the basis of our reciprocal dealings with our fellow human agents, and other individuals with whom we interact; and there is another kind of knowledge that we typically arrive at on the basis of our dealings with ''insensate'' physical objects, with which our relations go only one way. We observe objects; objects do not observe us in return; least of all—unlike our fellow humans—do such objects observe us observing them.

Given this basic need to contrast our knowledge of agents or persons with our knowledge of objects or things, one should expect a science of agents or persons to adopt somewhat different aims and methods from a science of things or objects. And, on a practical level, we are certainly accustomed to thinking about ''agents'' and about "objects" in rather different terms. The colloquial language in which we characterize our everyday experience itself reflects this fact, in a number of ways.

> On the one hand, the measure of our knowledge of objects is the extent to which we understand those objects (and the phenomena in which they are involved) and are able, on the basis of this understanding, to expect this or that to happen on given conditions. We come to know about objects; and this knowledge is true in the sense of *veracious*.

> On the other hand, the measure of our knowledge of agents is the extent of our understandings with those other agents (that is, our understandings about the courses of action in which they are engaged) and is expressed in the behavior that we are entitled, on the basis of those mutual understandings, to expect of the agents concerned. We come not just to know about our fellow agents, but to know them; and this knowledge is true in the sense of *trustworthy* or *reliable*.

Notice how the word "truth" links the idea of veracity with that of troth or trust. A true friend is different (how different!) from a truth-speaking friend, or a friend about whom we know the truth. A true friend is not just a friend about whom we can make veracious predictions, but one with whom we are linked by reciprocal hopes and reliances. The veracity or correctness at which the onlooker aims rests on the accuracy and scope of the expectations he has about things, and his understanding of them: it is the one-way kind of knowledge possessed by a spectator, who looks on (as it were) from a hide and is not, like a participant, reciprocally involved in the action. By contrast, the fidelity or reliability at which an agent aims rests on the depth and reciprocity of the expectations he has of other agents, and of the understandings that he has with them: it is the reciprocal kind of knowledge possessed by a participant who cannot detach himself from the action and view it from outside, like a mere spectator.

In the split world of post-Renaissance natural philosophy, the separation of "mind" from "matter," "reason" from "emotion"—and so humanity from nature—put a premium on the "objective" knowledge of the Spectator. It looked forward to the development of sciences whose results would be both intersubjective, that is, free from any personal slant, and also veracious:

sciences that would treat the entire world of nature as being itself an "object" about which the human mind could hope to reach perfectly accurate expectations and an exhaustively comprehensive theoretical understanding, free from irrelevant personal hopes and reliances. The classic expression of this idea was the familiar image developed at the very beginning of the nineteenth century by the great French astronomer and mathematician, Pierre Simon, Marquis de Laplace. If a scientist could only know all that God himself knew about the initial state of the physical universe at the Creation—if he only had at his disposal the initial positions and velocities of all the individual atoms in the universe at the moment of Creation—he would then (Laplace argued) be able, in principle, to apply the laws of motion and calculate the entire subsequent history of the universe. In this way, the intellectual ideal of Science which Laplace had inherited from Descartes and Newton was captured in the image of an Omniscient Calculator, whose grasp of the laws of science is manifested in the power to give a comprehensive and perfectly accurate running commentary on the entire historical action of the universe, *but from outside it.*

This last point needs to be underlined. In order for Laplace's image to characterize the "ideal standpoint" of natural science truly, it must be possible (at least in principle) for the scientist to observe, analyze, describe, and comment on the happenings in the world of nature *without being drawn into them.* The scientist must be able, that is, to remain in the position of William Shakespeare's Rosencrantz and Guildenstern rather than Tom Stoppard's, or of the drama critics in act 1 of *The Real Inspector Hound* rather than in act 3. He must be able, in real life, to adopt the standpoint that Descartes had conceived of for the human mind, as an abstract philosophical doctrine—set off apart from, and free from reciprocal interactions with, the material world of nature which is the *object* of observation and analysis.

So, the attitude of detachment and objectivity characteristic of natural philosophy in post-Renaissance Europe did not call merely for a psychological distance, noninvolvement, or decathexis: on the contrary, the scientist who hopes to be engaged and productive had better care about his subject matter, and preferably be passionately interested in the problems he chooses to work on. No: that attitude of self-conscious objectivity had deeper, philosophical roots. Most immediately, its ideal of rational objective knowledge sprang from the alliance of seventeenth-century science with Cartesian dualism; but its intellectual connections also went much further back, being rooted in the long-standing commitment of classical philosophy to *theoria*, the standpoint of the intellectual spectator.

How did the modern scientific movement's commitment to the "objectivity" of the spectator contribute, specifically, to the breach between science and natural theology? Suppose that we recall the contrast between knowledge of agents and knowledge of objects. Both kinds of knowledge have familiar parts to play in our everyday experience, and we might quite properly inquire just where the search for knowledge of each kind runs up against the limits of its natural scope, in what ways the two kinds of knowledge bear on one another, and how we are to relate them to one another in borderline situations. It would seem proper enough, that is to say, to accord both kinds of knowledge equal recognition, and then inquire on what terms and conditions we can maintain diplomatic relations with them both. From this point of view, there is no obvious reason why one kind of knowledge need claim intellectual superiority over the other: knowledge of objects over knowledge of persons, or vice versa. A decent respect for the scope and limits of each kind of knowledge is surely compatible with conceding the legitimate claims of the other.

That is not how things worked out in the actual history of philosophy from Descartes on. From the very start, Descartes's philosophical program had imperial ambitions. As he saw it, our understanding of geometry is not just one kind of understanding among others: it is intrinsically superior to other sorts of knowledge, and provides a model to which other branches of learning should aspire, to the extent that their basic concepts can approach the "clarity and distinctness" of the basic geometrical ideas and axioms. The resulting program for natural philosophy, of course, elevated sciences like physics, which are concerned with universal, general phenomena, to a higher level of respect and recognition than those other branches of learning that deal with particular, timebound individuals and events. The study of history, for instance, Descartes regarded as intellectually shallow. He likened it to foreign travel: it could at best broaden our experience, but could never deepen our understanding of basic principles—since, by the term "principles," Descartes meant only universal, timeless principles of a geometrical or quasi-geometrical sort.

Instead of viewing knowledge of agents and knowledge of objects impartially or evenhandedly, Descartes argued that the objectivity (that is, universality) of mathematical physics and similar sciences makes their kinds of knowledge more basic, more scientific, and even more *rational*: our knowledge of individual persons or agents, by contrast, can never be generalized or universalized *more geometrico* (in the manner of geometry) and is therefore superficial, unscientific, and subjective. This Cartesian commitment to the intellectual superiority of objective, universalizable knowledge, has shaped much of the subsequent development of the natural sciences. Even today, it remains strongly influential: for instance, in Noam Chomsky's scorn for anthropological linguistics, which he regards as intellectually shallow by comparison with the general theories of transformational grammar, and in those bitter methodological rivalries over the respective claims of "scientific objectivity" and "personal understanding" which split (and sometimes destroy) departments of psychology in American universities.

For our own purposes, however, we shall do better to stand back and look at this issue more coolly. The touchstone question remains the question whether one or the other kind of knowledge—knowledge of objects or knowledge of agents—can be *in itself* more basic, more scientific, or more "rational" than the other. How we answer that question will depend, first and foremost, on the direction from which we approach it. Suppose, in the first place, that we approach this question with the psychological development of the individual human observer in mind. There is little doubt that, in the lifetime of the individual child, knowledge of agents or persons is—in point of time and in depth of experience—the primary mode. Initially, the infant establishes a reciprocal understanding and familiarity with the significant human figures in its immediate milieu. Knowledge of Mother is not, at first, differentiated into distinct "cognitive" and "affective" components: factual expectations about mother-the-object on the one hand, mutual hopes, understandings, and reliances shared with mother-the-person on the other. (How many of us, indeed, can ever fully separate our expectations *about* our "nearest and dearest" from our expectations *of* them? For most of us, such knowledge remains a compound of accuracy and reliance, foresight and trust.) Developmentally speaking, therefore, any ability we may develop to set ourselves apart from our objects of knowledge, or to view them with the detachment of a pure onlooker, is secondary and derivative: it is an art that we learn only after a time, as we go along. That is why the onlooker's detachment involves a kind of abstraction. We begin our lives by being emotionally involved, and caught up in reciprocal relations, with all that we know at all well or closely: we can achieve the peculiar objectivity of the intellectual spectator only by learning to detach (or abstract) ourselves from that reciprocal involvement with the objects of our knowledge.

In the second place, suppose now that we approach the same question about knowledge of objects and agents, rather, from a more narrowly philosophical direction. In that alternative case, we may be tempted to set matters of developmental psychology aside, as being off the point. How the young child in fact develops has nothing to do with the important issues: we all know that young children are only partially and intermittently creatures of reason, and that they develop the capacity for dealing with the world in rational, intellectual ways only bit by bit, as they grow up. If we come from this alternative direction, accordingly, it will be natural enough for us to regard rational maturity as requiring an ability to view the world from an onlooker's detached point of view: that is, to abstract our knowledge and expectations *about* things in the world from all our personal concern *with* those things, and so to decenter our point of view toward them. This alternative line of approach may, thus, make us more sympathetic toward the traditional claims of scientific objectivity. True, when we encounter any theory that elevates the ability to view *persons* as *things* into a mark of maturity, a nagging doubt must remain about its complete adequacy: any such view has something bizarre about it. All the same, once we take the first few steps down Descartes's road, we shall quickly appreciate how the ability to abstract yourself from your objects of inquiry, and to view them from a detached standpoint, became for him and his successors a precondition for any reliable knowledge about them—and, still more, a precondition for any rational understanding of their place in the World-Machine.

Granted this initial detachment, the separation of natural science from natural theology followed easily enough. If the scientist merely has to observe nature, so that it is no part of his responsibility to interact reciprocally with nature, his ethical position is considerably simplified. If he is only looking on from outside and reporting what he sees, not acting in ways that produce significant effects from within the World, he can the more easily claim that his involvement with nature is morally neutral—that he is concerned only with ''facts'' and is not professionally concerned with values. Once the standpoint of the intellectual spectator is accepted as the only legitimate posture for the sciences (in short) a much stronger case can apparently be made out for imposing a hard-and-fast fact/value dichotomy on our scientific inquiries, for insisting on the value neutrality of science, and all that follows from that.

That done, it will be so much the worse for the kinds of universal interrelatedness that were the traditional concern of cosmology, such as John Donne's ''just supply'' and ''Relation.'' For those are certainly the kinds of relations that have *moral* significance: natural status, for instance, is not determined by geometrical coordinates alone; rather it has to do with the place that human beings *ought to* adopt vis-à-vis their fellow creatures. By itself, therefore, the fact/value dichotomy ends by cutting off science with an ax from all large-scale cosmology having any ethical or religious significance. Furthermore, once that starting point is granted, there can be no objection to splitting up science into as many subdisciplinary fractions as their specialized tasks require; the only facts to be reported about the whole of nature will then be those that are arrived at by aggregating the facts about its separate subdisciplinary aspects. Yet the question is: should we ever have conceded that starting point in the first place?

For some three hundred years, the natural sciences developed swiftly and fruitfully; and they did so, in part, simply because they adopted this novel Cartesian posture and program. For many previously neglected aspects of the natural world turned out, in fact, to lend themselves to fruitful investigation from a spectator's detached position. All those things whose behavior is in no way affected by the fact that they are under our observation can be studied and thought about with impunity as ''objects,'' for the purposes of science. So long as one is in fact dealing with ''objects'' (in other words) the pursuit of the spectator's ''objectivity'' remains a legitimate goal. As a result, the physical sciences moved ahead quickly during the eighteenth and nineteenth centuries, by dint of abstracting out for study just those aspects of nature which answered best to the demands of this "objectivity": planets moving around the sun, expanding and contracting gases, simple chemical reactions, and insensate rigid bodies or material objects such as balls rolling down inclined planes.

As seen from outside science, the scope of the resulting theories may well appear very narrow. For instance, the Cartesian separation of humanity from nature put serious obstacles in the way of any "scientific" psychology, at least until after the time of Kant; and there were serious problems even about simpler kinds of living things. (Historically speaking, vitality and "vitalism" have generated problems almost as intractable as those of mentality and "mentalism.") As seen from inside science, however, the things achieved within that limited scope appeared spectacular out of proportion to the results of any previous system of natural philosophy. So, it is no cause for surprise or shame that the methodology of Cartesian objectivity eventually became, in Francis Bacon's terminology, an *Idol*: that is to say, a way of thinking and arguing whose very power tempted people to press it beyond its own proper limits, and so to deceive themselves.

Even before the twentieth century, in fact, there were grounds for recognizing that the methodology of objective detachment and the associated standpoint of the intellectual spectator could not and should not be universalized. Even if one relied unquestioningly on the "classical" nineteenth-century scientific system, as built around Newton's mechanics and Maxwell's electromagnetism, any attempt to construct a fully comprehensive world picture must run up against these limits. For the situation of Laplace's Omniscient Calculator vis-à-vis the world of nature, which served to define the ideal standpoint for science, was one of absolute inequality. The Calculator's knowledge about the objects that he studied was complete, their knowledge about the Calculator was zero, so the "coupling" between them (so to say) went entirely one way. As a result, all the cards were in his hands, and his powers of prediction were unrestricted. By contrast, in the actual world, real-life scientists might approximate that situation quite closely vis-à-vis the planetary system and such like simple physical, chemical, and biological systems, but in many other situations—notably, vis-à-vis their fellow humans—they could not even hope to come close. In psychological research, the "coupling" between the scientist and his research subjects rarely goes entirely one way. A social scientist may approach this situation with humans by conducting a visual survey of pedestrians crossing the street at an intersection, by watching them from an upper-story window in a neighboring office block. But even that methodology is not infallible: if only those same pedestrians become aware of the social scientist's observing eye, we all know how countersuggestibly they may respond!

Even within nineteenth-century classical science, therefore, the two-way coupling between the observer and the observed already put philosophical limitations on the scope of the spectator's standpoint; and—as Karl Popper has argued—if those limitations had been more clearly brought to light at the time, that might have done something to counter the narrowly deterministic interpretations of the classical scientific world picture which were widespread during those years. During the twentieth century, however, similar limitations have become widely apparent even within the natural sciences proper. On one front after another, natural scientists have run up against the limits of the Cartesian methodology for themselves: indeed, this is perhaps the most significant general intellectual feature of twentieth-century science. Having mapped the scope of "detached, objective knowledge" from within (so to say) they have now reached its boundaries in a dozen different directions.

Some of the most striking illustrations of this novel development can be recalled on the level either of theory or of practice: whether we consider science in pure or applied terms. To begin with the theories of pure science: there the new limitations are apparent on every scale of inquiry. First, at one extreme, the finest-grain physical analysis of matter has come up against the limits expressed in Werner Heisenberg's "principle of indeterminacy." The precise significance of that principle for epistemology is still a matter for debate, but this much about it is clear. On the minutest level of scientific analysis, Laplace's ideal of the scientist, as "an unobserved, uninfluencing observer" studying the world of nature through a one-way mirror, is unattainable in principle for reasons of basic physical theory. There can be no simple, one-way coupling between a physicist and (say) the electrons that he selects as his objects of study. However delicate and miniscule our acts of observation on any subatomic particle may be, they will alter the particle's existing position or momentum, and so limit the precision with which its current condition can be known.

Second, on the everyday human scale, the limitations are more obvious. Now that human beings have unchallengeably entered the subject matter for scientific study, the objections to absolute detachment cease to be merely philosophical, and create active problems for science proper. The coupling between (say) psychoanalyst and analysand is evidently a two-way affair: the clinical problems connected with the phenomenon of countertransference, for instance, testify to the complexities that the reciprocity of this relationship introduces into the analytic situation. And that is, surely, as it should be: the understanding that a psychiatrist or psychoanalyst arrives at, as a result of dealing with a patient, will be none the better if he decides to regard the patient merely as an ''object'' and so rejects all the insights and understandings available from reciprocal interactions with the individual in question. In that respect, we can scarcely afford to base our scientific understanding of human psychology solely on data available to us from the Cartesian standpoint of intellectual spectators; nor can we afford to ignore the discoveries that come from dealing with our fellows in a reciprocal, participant's manner, and establishing understandings with them. (Calling to mind Kant's ethical maxim, enjoining us against treating our fellow humans as ''means only,'' we might even ask whether insisting on a spectator's objectivity in the study of human behavior may be not merely intellectually impoverishing, but downright immoral. Even research subjects are, and must be treated as, subjects, not mere objects!)

Finally, when natural science moves beyond the traditional Cartesian dichotomies, and reinserts humanity itself into the overall scientific world picture, the limitations of the spectator's posture once again become clearly apparent. The complex relationships that are the concern of ecology, for instance, involve human beings as elements within networks and cycles of natural interactions—in food chains, predator-prey, and other such two-way relationships—and the investigating scientist can no more opt out from these reciprocal relationships than can the farmer. Any ecological experiment, indeed, is itself not merely an observation but an action; and its effects on nature set the experimenter about as far from the detached situation of the undetectable Laplacean spectator as they possibly could.

Nor is this change in the status of scientific experiments, from observations to actions, merely a matter for philosophers. Instead, it is becoming a matter of practical, even political importance. In the controversy about recombinant DNA research, for instance, some of the natural scientists who most firmly resisted public supervision of the scientific enterprise rested their arguments on absolute and unvarnished appeals to the scientist's rights under the First Amendment of the United States Constitution: any State action to restrict freedom of thought and inquiry would be, they claimed, a violation of that Constitution. Yet, to take up this position absolutely (as Hans Jonas points out) is to argue as though a scientific experiment today were still a piece of mere "spectating," rather than being an action performed by a participant in the real world, with actual and possible effects on both Nature and the rest of Humanity. While the First Amendment arguments, and their underlying basis in common law, are certainly powerful and important, the presumptions they establish are far from irrebuttable; and, in any event, the need for some degree of public regulation of research was conceded many years since in connection with the handling of infective agents, poisons, and radioactive substances.

Similar tendencies are even more evident on the level of practice and "applied science." Before the year 1900, the range of practical activities that owed anything much to science was still quite small: electrical generation and transmission, the dyestuffs industry, and the telegraph were almost the only nineteenth-century industries heavily dependent on the results of scientific research. In fact, the major social transformations often referred to nowadays, collectively, by the use of the term "modernization," and attributed by some writers to the rise of "the scientific-technological society" (notably, population growth, urbanization, and the development of the manufacturing system) had already been underway well before the rise of modern science, and continued to develop for a long time independently of it. During the twentieth century, however, the coupling between scientific research and industrial technology has become much closer. The revolutions in communications from Marconi and Bell on; the upsurge in psychotherapy and psychiatry at the hands of Freud and his successors; the use of antibiotics in medical practice; the development of cybernetics and information processing beginning with Shannon and Wiener, and so on, and so on; it would be tedious to catalog at length the respects in which, from the early twentieth century on, the work of science has at last come to affect the life and practices of humanity, and the rest of nature also.

Furthermore, these changes in the role of science are taking place at the very time when other aspects of modernization—population growth, industrialization, and the rest—are themselves beginning to press humanity up against the limits of nature. Ours is the time when the problems of natural resources and energy utilization, environmental insults and endangered species—all those problems on which the ecology movement is focusing attention today—have ceased to be merely transient and local, and have become continuing and worldwide problems. In the process, the activities of human beings, their scientific activities as much as any other, have at last become significant elements within the operations of nature: in certain cases, critical and controlling elements. Far from being free to sit in the stands and watch the action with official detachment, like the original *theoroi* at the classical Greek games, scientists today find themselves down in the dust of the arena, deeply involved in the actual proceedings. They had thought of themselves as spectators; but they have been forced to double, at the very least, as team trainers and physicians.

To put our provisional conclusion in a single phrase: the scientist as spectator is dead. When Descartes and his successors set humanity over against nature, and converted the natural world itself into a mere "thing" or "object" (a *Gegenstand*, to use the revealing German word), they created a standpoint that was immensely fruitful to natural philosophers, for just so long as the work of science remained securely within the natural boundaries implicit in its definition. The option of viewing the world of things from an onlooker's point of view, and developing scientific theories embodying our "knowledge about" and "understanding of" those things, was, however, open to scientists only for so long as they confined their inquiries to things and objects, systems and phenomena, of kinds that lend themselves to study in that way. Now that we have discovered where the boundaries of that territory lie, scientists can continue pressing on beyond them as vigorously as before only if they are ready to modify their earlier ideal of "rational objectivity" and develop new methods of inquiry and patterns of thought—even, new *criteria* of objectivity and rationality—for use in these new fields of study. From now on, the scientists' conception of "theory" can no longer be that of pure spectator-dom: Laplace's ideal of the Omniscient Calculator has failed us, even in the purest and most fundamental parts of physics.

So Tom Stoppard is right. The posture of "pure spectator" is no longer open to us in natural science or philosophy, any more than it has ever been in social and political affairs. To insist on the value neutrality of science—notably, to insist on subordinating psychology and other human disciplines to the methodology of nineteenth-century physical science—is to turn the rational objectivity of the intellectual spectator into an Idol. There is, of course, some temptation to do just that. One of the traditional attractions of scientific work certainly seems to have been its apparent freedom from ethical ambiguities: if the search for pure scientific knowledge were self-justifying, the work of science would leave much less room for ethical conflicts and other human quandaries than there is in most professions, and the enterprise of science could legitimately be pursued in a withdrawn, monastic manner. This finally ceased to be possible, however, sometime during the earlier part of our century: at Hiroshima, if not before. By now, our position is very different; and we are having to take stock of all the assumptions that need to be reconsidered, if the traditional "scientific attitude" of Newtonian and Cartesian thought is finally to be set behind us.

At this point, I may allow myself an autobiographical remark. At the end of World War II, my philosophical curiosity led me to turn from physics and scientific cosmology—which had been my first loves, and by which I had first earned a living—and I went back to Cambridge University to "read Moral Sciences" as a graduate student. One small thing immediately puzzled me. Throughout the universities of Britain, the history of philosophy was divided up and taught under three headings: Ancient, Medieval, and Modern. Modern philosophy began with René Descartes, and in most places it scarcely went beyond Immanuel Kant. Yet, given my own scientific background and interests, I knew very well that, by almost any standards, the broad framework of seventeenth- and eighteenth-century ideas about nature from within which these "modern" philosophers had worked was more like that of Plato, Aristotle, and Democritus than it was like that of (say) Einstein, Schrödinger, and Sherrington—or even that of Darwin and Maxwell. So how could the philosophy of Descartes, Locke, or even

Kant, be properly labelled as "modern"? I was not mollified by being introduced to something called "contemporary" philosophy: especially because, at the hands of Bertrand Russell and G. E. Moore, this twentieth-century discussion so patently ignored the new intellectual framework of contemporary natural science, and instead persevered in the epistemological posture of the eighteenth-century British empiricists.

This puzzlement of mine might, of course, have extended more widely. When Herbert Butterfield began encouraging his academic colleagues in Britain to pay more attention to the history of science, his own ground-breaking book on *The Origins of Modern Science* displayed a similar oddity. It began in fourteenth-century Paris with Buridan and Oresme, and it hardly went beyond Isaac Newton. As Butterfield saw it, the fundamental change in people's ways of thought involved in the transition from the medieval to the modern world was the change that installed the methods and concepts of Newtonian physics in place of their Aristotelian forerunners. And those "classical" methods and concepts are precisely the ones which (as we now see) can no longer claim to be dominant and authoritative, whether in natural science, in philosophy, or in theology.

In short: if we are irremediably stuck with the existing academic divisions of historical time, we must reconcile ourselves to a paradoxical-sounding thought: namely, the thought that *we no longer live in the "modern" world*. The "modern" world is now a thing of the past. Our own natural science today is no longer "modern" science. Instead (to borrow a useful phrase from Frederick Ferré) it is rapidly engaged in becoming "postmodern" science: the science of the "postmodern" world, of "postnationalist" politics and "postindustrial" society—the world that has not yet discovered how to define itself in terms of what it *is*, but only in terms of what it has *just-now-ceased to be*. In due course, the change from modern to postmodern science will evidently be matched by corresponding changes in philosophy and theology also; in particular, the "postmodern" positions and methods that natural scientists are now working out will have implications, also, for a possible reunion of natural science with natural theology. What those implications are, we shall now turn and see.

The Fire and the Rose

Science and natural religion parted company (I have argued) for reasons that operated powerfully in the nineteenth and early twentieth centuries, but that no longer have the same power today. As a result of professionalization, the work of science was divided up between independent specialized disciplines, so that the broader concerns of traditional cosmology and natural theology ceased to be any scientist's professional business. Meanwhile, the role of the scientist came to be seen as that of a pure spectator whose task was simply to report "objectively" on the workings of nature; this perception carried with it a belief in the "value neutrality" of science, so that the moral, practical, and religious significance of our world view ceased to be a question on which science itself could throw any light.

All that has gone by the board during the present century. Within our own "postmodern" world, the pure scientist's traditional posture as *theoros*, or spectator, can no longer be maintained: we are always—and inescapably—participants or agents as well. Meanwhile, the expansion of scientific inquiry into the human realm is compelling us to abandon the Cartesian dichotomies and look for ways of "reinserting" humanity into the world of nature. Instead of viewing the world of nature as onlookers from outside, we now have to understand how our own human life and activities operate as elements within the world of nature. So we must develop a more coordinated view of the world, embracing both the world of nature and the world of humanity—a view capable of integrating, not merely aggregating our scientific understanding, and capable of doing so *with practice in view*. Only a broader, more coordinated view of the world of this kind can pick up once again the legitimate tasks undertaken by the traditional cosmology before the "new philosophers" of the seventeenth century led to its dismantlement.

Still, a price will have to be paid for this return to a richer view of the scientist's position vis-à-vis the natural world. Among other things, it means that the scientific enterprise can no longer be single-mindedly directed at the pursuit of pure knowledge, as it was so fruitfully for so long, between 1650 and the mid-1900s. Nowadays, scientists have always to consider themselves as agents, not merely observers, and ask about the moral significance of the actions that comprise even the very doing of science. In this respect, scientists today are having to rejoin the rest of humanity: their own professional work involves them once again in the moral and religious quandaries that arise for us all, in our attempts to reconcile action and reflection, the *vita activa* with the *vita contemplativa*.

This conflict between the values of contemplation and action is captured in two powerful images, around which T. S. Eliot constructed "Little Gidding," the third of his *Four Quartets*. The Rose represents the aesthetic, contemplative life. Its rewards are undeniable and even seductive, but transitory:

> Ash on an old man's sleeve
> Is all the ash the burnt roses leave.

In contrast, Fire represents the moral, active life. Its fruits are too often filled with pain, but it is driven by the power of Love, and holds out the promise of permanence:

> Love is the unfamiliar Name
> Behind the hands that wove
> The intolerable shirt of flame
> Which human power cannot remove.
> We only live, only suspire
> Consumed by either fire or fire.

Eliot's appeal to Love here parallels Heidegger's discussion of *Sorge*, or "concern." Both men alike—Heidegger as ever stolidly, Eliot in his own more sensitive but somewhat brittle manner—reject the purely contemplative life as incomplete. The *contemplator*, the *theoros*, the purely reflective and aesthetic spectator, pursues a genuine good; but, in the last resort, he is free to sit on his hands. The person who is moved by "concern," by contrast, may be fired (even "burned up") by the impulsion to do something about the objects of his concern; but the fruits of his mission transcend those of passive contemplation. That active, moral mission is now an essential element in the mission of science also, since this can no longer be directed solely at the classical kind of *theoria* or *contemplatio*.

The new mission and world view of "postmodern" science, as I have suggested, integrate our understanding of humanity and nature *with practice in view*; and this phrase requires some explanation. As we saw, John Donne recognized that the corrosive effects of the new philosophy consisted not merely in the fact that it dismantled Ptolemaic astronomy, but that in doing so it destroyed the "cohaerance" of traditional world view, and undercut the sense of "just supply" and of "Relation"—that is, the sense that human beings, like all other things, have a natural status within an overall scheme of things.

Certainly, Donne's words carried a moral burden. He was concerned not merely with what we can in point of fact believe about the natural world, but quite as much with our "place" or "station" in the world of natural things, and so with how we ought to live within the natural world. Deprived of their familiar roots within the previous orderly scheme of the cosmos, Donne's contemporaries were (it seems) swept away by a novel and unchecked moral individualism. The collective humanism of the sixteenth century gave way, in the early years of the seventeenth, to an individualism—even, a narcissism—familiar enough to us in our own time. Every man of spirit (Donne tells us) felt bound to prove that he was the unique representative of a unique species, owing nothing to the wreckage of the former cosmos:

> 'Tis all in peeces, all cohaerance gone;
> All just supply, and all Relation:
> Prince, Subject, Father, Sonne, are things forgot,

> For every man alone thinkes he hath got
> To be a Phoenix, and that then can bee
> None of that kinde, of which he is, but hee.

This passage of Donne's has complex implications, and bears thinking on very carefully. Not least, it paints a picture of the psychological background against which to reconsider the links between Descartes's personal attitudes and his philosophical thought. Among all the grandiose individualists of early seventeenth-century intellectual life, none was surely more of a Phoenix, both in his life and in his philosophical claims, than René Descartes; and one could make out a case for seeing him as engaged less in formulating a brand-new philosophical project than in finding new roots for a rather traditional view—not in the outside world, which had failed us, but rather inward, in the Ego.

Donne's argument, accordingly, wove together several different strands. Yet there was no necessary connection between those strands: in particular, between astronomy and the idea of "natural status." There was (that is to say) no inescapable reason why humanity should have measured its place in the traditional scheme of things by astronomical standards, rather than by others; and, if we set about reconstructing an account of "humanity's place in nature" today, we need no longer give astronomy the same dominant place that it had in pre-Renaissance cosmology. Indeed, since the year 1650, half a dozen things have helped to shift cosmological attention away from the heavens, and we can now look around for other areas of cosmological concern, even within the natural sciences.

The eighteenth- and nineteenth-century expansions in the scales of cosmic space and time, in particular—with the six thousand years of biblical chronology giving way to a cosmic time scale of more like ten thousand million years, and an even vaster increase in the spatial dimensions of the known universe—have swept away any assumption that the drama of human life on earth is playing itself out within a compact proscenium, marked out by the planetary system and the "sphere of fixed stars" supposedly adjacent to it. As it now seems, the stars have nothing particular to do with us. So, from the beginning, the fact that Newton's new theory of the heavens undermined the popular belief that astronomical phenom-

ena such as the appearances of comets were of immediate significance for human beings (i.e. "omens") struck his friend Edmund
Halley, and many others, as one of its greatest *merits*:

> . . . Nor longer need we quail
> Beneath appearances of bearded stars.

When we think about our own place in nature today, as a result,
astronomical matters may have some general relevance, but they
can no longer dominate our cosmology, as they did that of the
ancient Babylonians and the medieval Europeans.

In one crucial respect, in fact, our own scientific world picture
differs from those of both antiquity and the seventeenth century.
The pre-Renaissance astrocosmology shared with the scientific
world picture of the seventeenth century a belief that, in all its main
features, the world of nature has displayed the same fixed basic
structure throughout its existence. For the classical Greeks, it had
this fixed structure because the universe is indeed a *cosmos*: that is,
a single, well ordered, and self-sustaining system. For the Europeans of the seventeenth century, this was rather because the
universe is God's Creation, and was fashioned to conform to his
specification and to obey his laws, at least throughout the present
dispensation. Both the pre-Renaissance cosmology, based on principles of universal order, and the seventeenth-century picture of
nature, as structured by the divine design, were ahistorical cosmologies. Or, more exactly, the only significant historical elements in
those two views of nature were the events by which that fixed
structure was initially established—whether at the hands of Plato's
Demiurge, or of the Hebrews' Yahweh. In neither case was there
any continuing, progressive historical development within the
order of Nature as originally established.

Our own contemporary scientific picture of the world is, by
contrast, a historical—more particularly, an evolutionary—picture, above all else. In this respect, it shares more with the natural
philosophy of the Epicureans than it does with the other major
systems of Greek antiquity; for, in the classical world, only the
Epicureans regarded history as having a fundamental, unidirectional significance within the operations of nature. If the catch-
words of the classical cosmology were "cosmos" and "order" and

those of the seventeenth-century world picture were "harmony" and "design," the central themes of our twentieth-century world picture are, accordingly, "evolution" and "adaptation." The world of nature is the place where, as members within the larger evolutionary scheme of things, human beings are "well adapted," and so "at home"; where they have reason to *feel* "at home"; or, at the very least, where they have the power to *make* themselves "at home." When John Wheeler, the physical cosmologist and theoretical astronomer, spoke at the Smithsonian Institution's five-hundredth birthday party for Copernicus, indeed, he chose as his title a phrase that echoes this traditional concern—"The Universe as a Home for Man." For physicists as much as for anyone else today, the program of cosmology thus has an intrinsic connection with the ideas of "natural status" and "home."

This sense of being "at home" in nature is by now inextricably bound up with our understanding of evolutionary history. Human beings, like all other living creatures on the earth, are the beneficiaries of history. Specifically, we are the beneficiaries of millions of years of earlier "adaptive" changes, or "adaptations." And, to the extent that our future conditions of life will be, increasingly, what we *make* of them, our fate within this historical scheme depends, also, on the adaptiveness of our behavior, and so on our adaptability: more specifically, on the use that we make of our intelligence in dealing with our place in nature. In seeking to understand our place in this historical scheme, therefore, we shall have to reflect more carefully on the character and demands of our own adaptation. By doing so, we shall be taking the first constructive step toward restoring a proper sense of the practical concerns of cosmology.

For the Babylonians around 1000 B.C., understanding the natural world aright meant, among other things, ordering aright human life and interactions with the powers of nature. It meant, for instance, farming in accordance with the natural cycle of the seasons, irrigating in a timely fashion, maintaining a sound calendar, navigating prudently, recognizing the proper times and occasions for the annual festivals celebrating the powers of nature— which were also, of course, divine powers—and so on. Seeing how many of those practical cycles reflected the annual sequence of astronomical appearances and disappearances, we may find their initial preoccupation with the heavens quite pardonable. Yet who can doubt that these same issues are still of importance for us today? Through failing to understand and respect the ways in which our own activities interlock with the operations of the natural world, we too can easily end by farming destructively, irrigating ill-advisedly, navigating recklessly, placing excessive demands on nature, and generally losing a proper sense of "just supply" and "Relation." And we can do so just at that moment in history when all our practical ways of thinking most urgently require a proper sense of "cohaerance" and "Relation."

In fact, some first movement toward a revival of "natural religion," and a reunion of science with "natural theology," is already underway, though not necessarily under explicitly theological colors. The traditional issues of natural religion are forcing themselves on public attention, though under other names. It will be easier to recognize these new movements for what they are if we examine their philosophical genealogy.

The last period in Western history when humanity and the rest of nature were clearly thought of as complementary elements within a single overall scheme, or cosmos, was late classical antiquity. The intellectual allegiance of the general educated public—that is, men of the world, such as Cicero and Marcus Aurelius, rather than professional scholars, such as Simplicius and Plotinus—was then divided between two systems of "popular philosophy": that of the Epicureans and that of the Stoics. The Epicurean philosophy taught its adherents to practice detachment: that is, to avoid being disturbed by the afflictions of life, whether they sprang from natural or from human sources. From this point of view, a happy human being was one who had mastered his reactions to life, and could therefore keep feeling right about things: the key Epicurean virtue was *ataraxia*, not letting yourself be upset, not letting things "get to you." So, the Epicurean philosophy was primarily an inward-looking philosophy, teaching that self-command, that is, command over one's own inner, psychical resources, was more valuable than outward power, or command over the outward physical resources of the world. The Stoic philosophy resulted in a very similar set of ethical maxims, but it arrived at them from the opposite direction.

It looked outward rather than inward: it taught its adherents to look for the sources of inner, human order and rationality in the external order of nature, and exhorted them to live in harmony with nature. By giving their own lives a *logos* that fitted harmoniously the *logos* of nature, they could—as far as was practicable for a human being—avoid being exposed to natural afflictions, and so achieve both personal tranquility, or *apatheia*, and also good-spiritedness, or *eudaimonia*.

Now that "postmodern" science is seeking to reinsert humanity into the world of nature, it should be no surprise to find that people are being drawn, once again, toward religious and philosophical ideas that are highly reminiscent of late antiquity. And, indeed, the "popular philosophies" current among the educated public today are reviving some strikingly Stoic and Epicurean themes. Two systems of ideas, surely, dominate nonacademic thinking about philosophy and ethics today: what may be called, for short, the "white" philosophy and the "green" philosophy.

The white philosophy has roots in psychotherapy. It encourages us, above all, to pursue self-knowledge and self-command as individuals. Our prime responsibility as human beings is to identify the points of inward frailty within our personal psyches—the elements in our childhood Pantheons which leave us vulnerable to aggravation by outside agents and events. We are then to master those frailties, by facing our inner "ghosts" and so exorcising them. In this way (it is claimed) we can make ourselves impervious to aggravation, and learn to prevent things from getting to us. As a result, we shall be able to control our own reactions, and stay cool, whatever happens outside us. In all these respects, the "white" philosophy—the philosophy of psychotherapy—recalls the Epicurean option of late antiquity.

The green philosophy, by contrast, is a contemporary counterpart of Stoicism. It has roots, most typically, in the theories of ecology and the practices of "natural living." It encourages us, both as individuals and in our collective affairs, to pursue harmony with nature. Our primary responsibility is to deepen our understanding of the interdependence that binds humanity to nature. We are then to direct our lives, on both the personal and the social levels, in ways that do not interfere with the cycles and systems of the natural world but go with the grain of nature. In this way (it is claimed) we can avoid subjecting nature to traumatic insults that will recoil on us, and so be sources of inescapable aggravation. As a result, we shall achieve a command over outside events based on mutual respect rather than domination, and so have every occasion to stay cool, whatever happens. In our relationship with nature, at least, we shall have clean hands.

In antiquity, the Stoic and Epicurean positions were philosophically at odds. The Stoics shared with Plato and Aristotle a belief that the fundamental structure of the "cosmos" is ahistorical; or, at most, that it changes in a cyclical, and so repetitive manner. The Epicurean system was a basically historical system; from the Epicurean point of view the "principles" of the natural world were no more general and fundamental than the unique, nonrepetitive unfolding ("evolution") of historical events through time. An initial disordered rush of the atoms through the void had been followed by a natural aggregation of material objects, then by the appearance of living creatures on the earth, and finally by the establishment of human societies. At each stage, some qualitatively new principles of organization entered into the operations of the world; and a philosophical preoccupation with unchanging or timeless principles alone would only distract one from the significance of historical innovation. Atoms and the void alone remained throughout the whole of cosmic history: everything of order and value—everything of genuine human concern—was a product of historical change. Preserved in the traditions of Magna Graecia and transmitted to the scholars of seventeenth-century Naples, the historical cosmology of the Epicureans was to surface again through the curious and equivocal pen of Giambattista Vico early in the eighteenth century.

The ecological ideas associated with today's "green philosophy" have, however, nothing ahistorical about them. On the contrary, ecology both learns much from the study of biological evolution, and contributes much to it. If we are to understand how the physical and mental attributes of the human species (or any other species) are "adapted to" the conditions of terrestrial life, we need to consider among other things how they came to be adapted as they are. This may be, in part, a story of "evolutionary mistakes'': for instance, zoologists associate many of our spinal and digestive problems with the fact that human beings have adopted a vertical posture, instead of continuing to go around on all fours. (Our spinal discs, for example, are ill adapted to the pressures resulting from this new, vertical posture.) But such mistakes or maladaptations are exceptions, rather than being the general rule. In countless other respects, human beings are as they are because it was well and still is well that they should be so. Since this alliance with evolutionary biology makes modern ecology a historical science, it also removes the chief obstacle to an alliance between the white and the green philosophies themselves. In our own times the preoccupations of the Stoics are more easily reconciled with those of the Epicureans than was possible in antiquity; and, certainly, plenty of people today in fact seek to combine an Epicurean trust in the insights of psychotherapy with a Stoic commitment to the virtues of natural living and appropriate technology.

From this combined point of view, it is once again possible to reunite the worlds of humanity and nature into a true "cosmos"— an evolutionary cosmos, to be sure, but a cosmos none the less. In two crucial and central respects, that is to say, postmodern science puts us in a position to reverse the cosmological destruction wrought by modern science, from A.D. 1600 on. The world view of contemporary, postmodern science is one in which practical and theoretical issues, contemplation and action, can no longer be separated; and it is one that gives us back the very unity, order, and sense of proportion—all the qualities embraced in the classical Greek term *cosmos*—that the philosophers of antiquity insisted on, and those of the Renaissance destroyed. So human meaning finds its way back into our picture of the natural world, from which it was banished in John Donne's time.

To recognize the cosmological point of these contemporary ecological and psychotherapeutic doctrines and attitudes is not necessarily to accept, or even to recommend them. Still, it would be dishonest to deny their interest and charm, quite apart from their historical significance. For the moment, however, let us simply remark that the popularity of these two neo-Hellenistic philosophies reflects a renewed interest in the practical concerns of cosmology and natural religion. As human beings, we need to understand our own position vis-à-vis the rest of nature, in ways that will permit us to recognize, and feel, that the world is our "home." In addition, we need to discover in what respects, and on what conditions, the world of nature can continue to provide a home for humanity. Only then can we learn to handle ourselves in such a way that we are truly at home in the natural world, and that the natural world itself is capable of remaining the kind of home it can be for human beings. Those, of course, have always been among the central tasks and themes for cosmology.

As sketched here, these popular views are, of course, somewhat unsophisticated and unanalyzed. Yet, at the very least, they give useful pointers toward the issues that will need to be addressed by any future "theology of nature," and toward the problems that must be analyzed if such a new cosmology is to stand up to criticism, and carry conviction, after three hundred and fifty years in limbo. How, then, can these issues and problems be characterized concisely, within the limitations of a single essay? Let us consider some representative samples.

Suppose, for instance, that we recognize the twin concerns of ecological thought and, for purposes of comparison, set "ecology" the pure, theoretical science, alongside "ecology" the social and practical philosophy. The natural science of ecology can be personified in Evelyn Hutchinson of Yale. Hutchinson has devoted his distinguished career as a biologist, in part, to creating a new disciplinary subbranch of ecology called "limnology," that is, the science of ponds. As he realized, compact, enclosed bodies of water provide ready-make laboratories in which we can study the vital processes of ecological stability and change in naturally controlled and predictable milieux. So he set himself to discovering the food chains, chemical balances, seasonal cycles, and other ecological processes typical of self-maintaining ponds. In his own purposely unobtrusive manner, Evelyn Hutchinson has thus improved our general grasp of the interactions and relationships with which the science of ecology is concerned, and has put us all in his intellectual debt.

The social philosophy of ecology can be personified, by contrast, in John Muir of San Francisco. Muir was the person whose concern for the California redwoods led him to launch the movement for nature conservation first embodied in private organizations like his own Sierra Club, and also now in public agencies like the Environmental Protection Agency. As he saw matters, the unthinking destruction of such natural systems as the California redwood forests raised moral, not just social and pragmatic issues. Aside from anything else, the redwoods did not deserve to be destroyed! As a result, John Muir "raised the consciousness" first of the lay public, and later of governments, about our responsibility for managing the balance between humanity and nature in an intelligent, sensitive, and discriminating manner; and he has put us all in his moral and political debt.

When we compare Evelyn Hutchinson and John Muir two points stand out. To begin with, ecology the pure science and ecology the practical philosophy are clearly different. If we were to perform a Cartesian act of abstraction and intellectual detachment, we could approach the phenomena of ecology from the standpoint of pure theory, focusing on the basic processes within ecological systems out of an intellectual concern with scientific principles, without regard to practical policies. If we insisted on it, likewise, we could approach the political problems of ecology from the standpoint of pure practice, devoting ourselves to campaigns on behalf of endangered trees or fish or plants out of a practical concern with political issues, without regard to scientific principles. Both these courses of action are possible, but neither of them is any longer easy, since both of them involve the kind of disciplinary abstraction and fact/value separation which no longer appears as legitimate as it did a hundred years ago. The scientific and political aspects of ecology may be distinct; but, in the age of postmodern science, they are not so easily kept separate.

Nor is there any significant conflict between them. Informed and practically discriminating ecological policies depend on accurate and intellectually discriminating ecological theories; while, in return, the practical policies and experience of (say) the Forestry Service provide a significant source of new material for ecological scientists. Ernst Mayr, for instance, has explained the importance of burning over public woodlands from time to time, instead of preserving them from international or accidental fire at all costs. It is not just that the phenomena associated with forest fires have scientific significance: in addition, the corresponding policies have practical value. We may be tempted to suppose that the best and most natural forest is necessarily a fire-free forest, but this is a misconception. The natural action of fire long antedated human carelessness. Lightning, for example, was starting forest fires long before matches were invented, and most natural forest species are in fact "adapted to" occasional fires. In some species of coniferous trees the fallen cones will actually remain tightly closed until they are roasted by a forest fire: only then can the seed escape from the cone, fall into the soil, and germinate. If the forests in which these trees grow never burned over, they would eventually become extinct.

As to the second point about ecological science and ecological philosophy: once these are seen in their true relationship, it becomes clear that, far from being purely "factual" in their content, many of our central biological concepts are also by implication *ethical* concepts. The human species is directly involved in a great many ecological processes, often as their prime agent or even as their victim; so these processes cannot be treated as ethically indifferent "objects," for detached theoretical study alone. Nor can the associated questions about the mutual adaptation of human beings and other living things be regarded as purely theoretical questions for a detached spectator. Rather, the very concept of adaptation has an inescapable ethical component. The central biological question is not just "How are human beings (passively) adapted to the natural habitat?": it is, also, "How are we (actively) to adapt ourselves to that habitat?"

We need to understand, not merely the relations that hold between humanity and nature in point of observed fact but also (as John Passmore puts it) the ethical considerations involved in "man's responsibility for nature." This means asking, not just "How, on what conditions, and with what effects, has the human species in fact coexisted with (say) the horse, the snail darter, and the smallpox virus?"; but, in addition, "How, on what conditions, and to what ends, ought the human species choose to continue coexisting with those other creatures?" And, when we begin to ask what it is that scientific intelligence and moral discrimination, between them, demand of us in our relations with the rest of nature, the dreams of natural philosophy become both the concerns of natural religion and the responsibilities of public policy as well.

This is not to claim that ecology and psychotherapy—either separately or together—by themselves provide the sole, or even the most legitimate, heirs and assigns of natural theology surviving today. All the same, the development of these novel fields illustrates well how thoroughly the older view of value-free natural science has broken down, and how much we now need to rethink our beliefs about the place of humanity in some larger scheme of things. Those beliefs were, of course, traditional concerns of cosmology and natural theology; and, now that scientists have abandoned the spectator's standpoint to these cosmological questions, they are beginning to arise again spontaneously, even within science itself. That being so, the question we posed at the outset— namely, how far contemporary science leaves room for natural religion, or a "theology of nature"—has answered itself. There is indeed room for scientists, philosophers, and theologians to sit down together and to reexamine in detail the scientific, ethical, and theological issues that arise about such ideas as "natural status" and "the larger scheme of things" (*cosmos*). At the very least, there is room to consider just how the associated scientific, ethical, and theological issues bear on one another.

How, then, do professional theologians react to this prospect? They do so with mixed feelings. To pick out two representative samples for comment, we may contrast a striking address at the University of Chicago by Hans Küng, the Catholic theologian from Tübingen, with a paper delivered by James Gustafson to a Hastings Center working group on "Science and the Foundations of Ethics."

In his address on "Science and the Problem of God," Dr. Küng shrewdly analyzed the fallacies that await us if we start from the limited concepts of any single scientific discipline, and then attempt to generate a comprehensive world view by generalizing them without limit. When we generalize the concepts of any one particular discipline in this way, he showed, we can too easily end by undercutting even the legitimate, but more limited, disciplinary claims those same concepts have on our intellectual allegiance. Generalizing Freudian concepts without limit, in the hope of generating a comprehensive "Freudism" and so undermining the claims of religion, finally recoils on the generalizer himself, by raising the question whether his own arguments are not themselves so many neurotic symptoms. Similarly, generalizing without limit the concepts of sociology or neurophysiology (say) can again recoil on the generalizer, by raising the question whether his own arguments are not themselves either the biased products of class interest or the effects of synaptic discharges in the cortex.

Küng's line of attack on some fashionable lines of antireligious argument is, thus, a highly destructive and effective piece of polemic. But, when the time comes to put forward constructive countersuggestions, his own exposition goes unaccountably lame. He seems to have no constructive program to offer for a reunion, or even a reconciliation, of science with theology. Instead, he is apparently content to preserve the existing armed truce. Least of all does he show any readiness, on his own account, to abandon the spectator's purely theoretical posture. Having shot off his bolts at the scientists, he simply retreats onto the high ground of traditional metaphysical thinking—the ground of total generality, onto which no levelheaded scientist would be so foolish as to follow him. Once there, he wraps himself in the ultimate cosmological mystery of the *Urgrund*: that is, the answer to the most theoretical question of all, "Why does the world contain *anything*, rather than *nothing*?"

James Gustafson's paper is more constructive, and also more specific. He sets out to list the features that mark off a specifically religious approach to ethics; and his final formulation comes close—in spirit, at least—to the outcome of our present line of argument. Adopting a religious attitude to ethics and experience (he says) means resolving among other things to deal with, and relate to, all created things in ways appropriate to their relations to God; for which one may, without too much distortion, read instead, "dealing with all our fellow creatures in ways appropriate to their places in the overall scheme of things." This formulation can open our eyes, and our minds, to the cosmological questions on which our own argument has also been focusing attention; and it has two further considerable merits for anyone who is interested in exploring the detailed theological affiliations of contemporary natural science.

Instead of retreating to the metaphysical snows, Gustafson comes to grips with specific, practical issues. He states his position in terms that have real practical consequences: inviting us not merely to think about (theorize about, adopt a spectator's attitude toward) the overall scheme of things, but also to treat (act toward, adopt a participant's attitude toward) all our fellow creatures in ways that respect their place in the natural scheme. Again, he states his position in terms specific enough to gear in effectively with our detailed scientific concerns and insights: inviting us to think of, and to treat, each kind of thing in the world in the specific ways appropriate to its particular natural status.

Appeals to the *Urgrund*, of course, do not enable us to discriminate between things of different sorts. The question "Why is there anything at all?" lumps such diverse questions as "Why are there human beings at all?" and "Why are there snail darters at all?" with "Why are there smallpox viruses at all?" and even "Why are there grains of sand (or radioactive substances) at all?" By contrast, the demand that we treat things in ways that reflect our perceptions of their specific places in some overall scheme enables us to adopt carefully differentiated attitudes toward moral claims on behalf of (say) snail darters, which we regard as valuable, deserving, or at least innocent, and smallpox viruses, toward which we are prepared to show no such mercy. If we insist on discussing "man's responsibility to nature" in entirely general terms, we eventually lose contact with the concrete reality of moral choice. In order to go any further, we must be prepared to harden up the debate, by addressing the specific responsibility of humanity toward snail darters, smallpox viruses, and other equally particular kinds of things. This sort of specificity is, surely, the very minimum we can require of any theology of nature that is to carry conviction today.

Is even that too much to ask? In these essays, we have defined some common ground for a renewed discussion between natural science and natural theology; but it remains to be seen whether the consequent debate will lead to fruitful results. Before launching ourselves too enthusiastically at such highly specific issues as the rights of snail darters, therefore, we must spend a little longer on the preliminary, transcendental question: namely, "On what conditions is a *fruitful* reunion between natural science and natural religion *überhaupt möglich*—that is, possible at all?"

If we reflect on this question for a moment, certain restrictions quickly become apparent. At the very best, science and natural theology will be able to renew their alliance fruitfully only if certain conditions are satisfied. By Gustafson's standards, for instance, only certain ecological approaches will be either theological in their intellectual affiliations or religious in their practical implications. At the present time, the ecological movement is itself divided into two groups: some of its supporters still have basically anthropocentric, even utilitarian attitudes, while for others Nature itself has become once again an "overall scheme" or cosmos, and so a target for genuine piety. In the one group there are those whose support for enlightened ecological policies rests ultimately on an appeal to our own longer-term interests as human beings. "If we only know what is good for us and our successors," their argument goes, "we shall take care not to deplete the earth's gene pool by allowing endangered species to become extinct, and so protect the interests of our human descendants, for whom the diversity of nature may provide a useful pool of instrumentalities." In the other group there are those others whose view of our place in nature is more genuinely cosmological: who find themselves compelled to set utilitarian, anthropocentric arguments aside and extend to other forms of life an honorary (though nonvoting) citizenship in Kant's Kingdom of Ends, together with the dignity and respect appropriate to that citizenship.

As to this division, only one thing needs to be said here. Those whose attitude toward "the larger scheme of doings" remains purely anthropocentric will, indeed, be failing to satisfy Gustafson's criterion: they will be "relating to other created things" in ways determined, not by their respective relations to God but rather by their relations to ourselves and to our own human interests. They will, that is, still be claiming for our species all rights and authority to deal with other kinds of things (notably other forms of life) merely according to our own convenience, will, and desire, rather than according to any properly established conception of their places in some overall "scheme of things." Yet the unqualified extrapolation of a utilitarianism that lacks any ultimate, intrinsic goods other than our own wishes and preferences quickly lands people in a sense of absurdity; the fashionable notion of absurdity being closely linked to the loss of any sense of natural place in the world. So, the attempt to vindicate ecological policies in anthropocentric terms alone has an unhappy destination. Let our human goals be as long-term and enlightened as they may, it will still be a case of humanity utilizing nature rather than respecting it: that is, treating all other creatures, not as ends in themselves, but rather as means only. By what final warrant? To what larger good, beyond human pleasure and convenience? Who can say?

There remains the other alternative. We can do our best to build up a conception of "the overall scheme of things" which draws as heavily as it can on the results of scientific study, informed by a genuine piety in all its attitudes toward creatures of other kinds: a piety that goes beyond the consideration of their usefulness to Humanity as instruments for the fulfillment of human ends. That is an alternative within which human beings can both *feel*, and also *be*, at home. For to be at home in the world of nature does not just mean finding out how to utilize nature economically and efficiently—home is not a hotel! It means making sense of the relations that human beings and other living things have toward the overall patterns of nature in ways that give us some sense of their proper relations to one another, to ourselves, and to the whole.

If we let ourselves accept anything less than this, we shall not yet have integrated our understanding of humanity and our understanding of nature in a way that "has practice in view." Our natural science and our theology of nature will still be going off in different directions. The demands of the *vita activa* and the *vita contemplativa* will still be at odds; and we shall not yet be at home. To return to Eliot's "Little Gidding": notice how this figure of "home" comes in at the very end of the poem, and provides him with a resolution of the divergences between contemplation and action, between the Rose and the Fire.

> We shall not cease from exploration
> And the end of all our exploring
> Will be to arrive where we started
> And know the place for the first time . . .

And all shall be well and
All manner of thing shall be well
When the tongues of flame are in-folded
Into the crowned knot of fire
And the fire and the rose are one.

It would be agreeable if we could simply stop at this point, with Eliot's lucid image in mind. But honesty compels us to muddy the waters a bit, and we are still left with far more questions to ask than questions answered. If there is to be a renewal of contacts between science and theology along the lines suggested here—if the cosmological presuppositions involved in talking about the "overall scheme of things" are to be scrutinized jointly from both sides of the fence—we shall quickly encounter some knotty problems of jurisdiction.

One of these problems revives in contemporary form a question that was already familiar in the Middle Ages. Like common morality, scientific ecology is presumably a field for exploration in the light of "the natural reason." To that extent, it should not be a field for doctrinal or denominational disagreements. Yet does this mean that scientific ecology, on its own, can tell us all that we need to know for ethical and political reasons, too, about the mutual relations between different living species within "the overall scheme of things"? Very early on in any such discussion, we shall have to face the question: "Just how far along the road to a theology of nature *can* scientific disciplines like ecology take us? And, beyond that point, what legitimate room exists for sectarian disagreement and doctrinal particularism?"

This is not just a speculative question. We have the strongest practical reasons for maximizing the area of rational agreement about "the overall scheme of things," as between people whose other convictions are quite different. The winds blow and the rivers flow as they do naturally, without respect for human differences over "matters of faith and doctrine." The same ecological problems that demand respect by Americans and Canadians for the fauna and flora of Lake Erie, or by Italians for the Venetian lagoon, are already affecting Lake Baikal in Soviet Siberia, and could arise tomorrow about Lake Chad or Lake Nyasa in central Africa. Any vision of "humanity's place in nature" effective on the level of international policy will, therefore, have to be presented in terms that are, at the very least, not offensive to "the natural reason."

Yet does this put us in a position to claim, quite baldly, that the entire scheme of Creation by which our moral and religious ideas are to be guided is transparent to "the natural reason" without regard for the doctrinal considerations of particular religions and sects? Preachers who exhort good Christians to let their Christianity permeate all their thinking, so that they may even end up with (say) a "Christian arithmetic," invite Leibniz's objection that arithmetic is just not like that—even God himself cannot alter, or contravene, the truths of mathematics. And, if we were told that good Christians must subscribe to a different science of ecology from other people, a parallel objection might well be pressed. God intervenes in the World (Leibniz declared) within the realm of grace, not within the realm of nature. So perhaps the time has come to take our courage in both hands, and declare for a fully common and ecumenical theology of nature. If we do so, we shall certainly not win universal support. However open we may find some liberal theologians, Catholic and others, to the idea that natural theology should be accessible to the common reason, we cannot hope for the same tolerance from all Protestants. The history of strained relations between scientific theory and fundamentalist theology is too long-standing for that.

Just how far, then, can the natural reason alone inform us in detail about what the overall scheme of things—the cosmos, or Creation—really is? Just how far can it tell us how we ought to act toward the other kinds of creatures that have their own proper places in that scheme—toward whooping cranes or smallpox viruses, toward sign-using chimpanzees or the fish of Lake Erie? These essays have not attempted to settle such questions as those: only to show that they are among the legitimate heirs of natural theology in our own day. Before they can be settled, scientists, theologians, and philosophers will have to sit down together and follow their joint discussion where it leads. We have reached the threshold of some painfully difficult and confusing questions, but answering them is a task for the future.

Index

Act of Creation, The (Koestler), 90–113, 177

Adaptation, 134–135, 143, 160, 161, 207, 208, 267

Allgemeine Natureschichte und Theorie des Hummels (Kant), 4

Almagest (Ptolemy), 10

Analysis of Matter, The (Russell), 212

"Anatomie of the World" (Donne), 219, 227

Anaxagoras, 9

Anaximander, 2, 172

Antic Hay (A. Huxley), 68

Apocalypse myths, 33, 35, 84; and scientific theory, 35–42, 47–49. *See also* Running-down universe

Archimedes, 95

Aristophanes, 172

Aristotle, 16–17, 54, 55, 109, 206, 238–239; *Metaphysics*, 150

Astrocosmology, 221, 222–227, 231; functions of, 225; themes of, 222–225

Astronomy, and cosmology, 4–5, 8–10, 221, 227. *See also* Astrocosmology

Atlas myth, 23, 69–70, 81

Babylonika (Berossos), 222

Bachelard, Gaston, 151

Bacon, Francis, 127

Bateson, Gregory, 15, 201–213; central mission of, 203–204; "double bind" theory of, 204; *Mind and Nature: A Necessary Unity*, 205–209, 210–213; *Steps to an Ecology of Mind*, 205

Bateson, Mary Catherine, 202

Bateson, William, 203

Behavior, and evolution, 133–134, 183, 203–205, 208–209

Behaviorist psychology, 181–183, 186

Berger, Peter, *The Structure of Scientific Revolutions*, 211

Bergson, Henri, 14, 99, 119, 149

Bernals, John Desmond, 230

Bernard, Claude, 151, 156, 187

Berosses, *Babylonika*, 222

Beyond Freedom and Dignity (Skinner), 182

Beyond Reductionism (Koestler, editor), 178

Biologists: French (*see* Jacob, François; Monod, Jacques; Teilhard de Chardin, Pierre); world views of, 75–77, 78

Biology: and ethics, 32, 56–70, 267; Koestler's views on, 102–104; molecular, 157–158, 160–161; structuralism in, 161–162, 164

Bisociation, Koestler's concept of, 94–101, 104, 111, 112, 177, 186

Blake, William, 187

Bohm, David, 13

Boltzmann, Ludwig, 160

Boyle, Robert, 232

Brady, Joseph, 182

Brain: evolution of, 135, 169–170, 175; psycho-pharmacology and, 199

Bridgwater Lectures, 218

Bridgwater Treatises, 128

Buffon, Georges-Louis Leclerc, *Epoques de la Nature*, 170

Burnet, Thomas, *Sacred History of the Earth*, 9

Burt, Cyril, 104

Butterfield, Herbert, 84; *The Origins of Modern Science*, 254

Caesar, Gaius Julius, 193
Callisthenes, 222
Cambridge Journal, 14
Camus, Albert, 143
Cartesian: dichotomies/dualism, 141, 212, 213, 224, 238, 243—250, 252—253, 255; mechanicism, 14—15, 111, 130—131, 147, 160, 161; methodologies, 130—137, 140, 209, 210; physiologists, 130—131, 157. *See also* Cartesianism
Cartesianism, 140, 156, 162, 164, 195; limitations of, 164, 209, 238, 248, 249, 250, 255, 266
Case of the Midwife Toad, The (Koestler), 178
Catholic theology, 115, 123—125
Ceres, 23, 70
Chance and Necessity (Monod), 141—155
Chardin. *See* Teilhard de Chardin, Pierre
Chemistry, atomistic concepts of, 9
Chomsky, Noam, 145, 148, 175, 245
Christianity, 123—125. *See also* Catholic theology
Christmas at Cold Comfort Farm (Gibbons), 124
Coleridge, Samuel Taylor, *Kubla Khan*, 97
College de France, 141, 157
Computers, electronic, 110, 111—112
Conjectures and Refutations (Popper), 127
Constitution, U.S., First Amendment to, 251
Copernican theory, versus Ptolmaic, 81—82, 227, 257
Copernicus, Nicolaus, 95, 132, 176, 179, 221; *De Revolutiombus*, 10—11
Cortical functions, 141. *See also* Neocortex
Cosmology: and astronomy, 4—5, 8—10, 221, 227 (*see also* Astrocosmology); and ethics, 53
Cosmos, Greek conception of, 224, 259

Creation: myths, 33, 34; physical theories of, 49—52; and scientific myths, 84; views of, 49—50, 51, 52
Creation, The (Haydn), 217
Creative process, Koestler's views on, 90—113
Crick, Francis, 157
Critique of Pure Reason (Kant), 4, 6, 9, 150
Cumont, Franz Valéry, 222
Cuvier, Georges Léopold, 156

Darkness at Noon (Koestler), 139, 177, 197
Darwin, Charles, 94, 95, 160; *The Descent of Man*, 165; on ethics, 61; *The Expression of the Emotions in Man and Animals*, 166—167, 203; *The Origin of Species*, 146, 165, 218; and psychology, 166. *See also* Evolution theory, Darwinian
Dawson, John, 14
Delbrück, Max, 157
De Revolutiombus (Copernicus), 10—11
Descartes, René, 3, 16, 29, 126, 162, 238, 258; *Dieu philosophique*, 224; natural philosophy of, 244—245, 246; view of solar system, 29—30. *See also* Cartesian; Cartesianism
Descent of Man, The (Darwin), 165
Detachment, scientific, 212—213, 237—253, 256
Dewey, John, 154
Dialectics of Nature (Engels), 2, 10, 17
Dichotomies. *See* Dualism, Cartesian
Die Ende Aller Dinger (Kant), 4
Dieu philosophique (Descartes), 224
Dingle, Herbert, 48
Dobhansky, Theodosius, 114
Donne, John, 220—221, 231, 238, 247, 257—258; "Anatomie of the World," 219, 227
Dragons of Eden, The: Speculations on the Evolution of Human Intelligence (Sagan), 168—175
Dualism, Cartesian, 141, 212, 213, 224, 238, 243—250, 252—253, 255
Duhem, Pierre, 11, 12, 151; *Physique d'un Croyant (The Physics of a Be-*

liever), 11, 115

Durkheim, Emille, 207

Durrell, Lawrence, *Mountolive*, 222–223

Ecology, science versus philosophy of, 265–267

Eddington, Arthur S., 73

Einstein, Albert, 15, 95, 151, 179

Eisley, Loren, 170

Eliot, T. S., 25; *Four Quartets*, "Little Gidding," 256–257, 272–273

Embryological development, Koestler on, 102, 103n

Empedocles, 9

Engels, Frederick, 141, 149; *Dialectics of Nature*, 2, 10, 17

Entropy, 22, 26, 28, 37, 40, 41, 42, 48. *See also* Running-down universe; Thermodynamics, Second Law of

Environmental Protection Agency, 266

Epicureans, 36, 167, 262–263, 264

Epoque de la Nature (Buffon), 170

Erickson, Erik, 202

Ethics: and biology, 56–70, 267; cosmological foundations for, 53; and evolution, J. Huxley on, 56–69, 74, 129; Sagan on, 175; and scientific myths, 84; of scientific work, 230; Spencer's views on, 71

Evolution: ambiguous use of term, 26; and ethics, 56–69, 74, 129 (*see also* Huxley, Julian); of brain, 169–170, 175; T. H. Huxley on, 57, 60; and scientific mythology, 22, 26–27, 58–70; and sociocultural change, 163; Spencer's view of, 121; Teilhard de Chardin's ideas on, 119–120. *See also* Evolution theory *headings*

Evolutionary Ethics. *See* Evolution, and ethics; Huxley, Julian

Evolution theory, Darwinian: dispute over, 10, 26–27, 194–195; Jacob on, 14, 156–161, 163; Monod on, 14, 141, 143, 144–146, 147, 152; Shaw on, 121; strength of, 202; Teilhard de Chardin on, 116–120,

124; variation in, 144, 145–146, 147

Evolution theory, Lamarckian, 14, 120, 121–122, 137, 140, 177, 184, 190

Evolution theory, neo-Darwinist, 145, 146, 153, 157; Koestler on, 109–110, 183–185, 190–191, 195–196

Expression of the Emotions in Man and Animals, The (Darwin), 166–167, 203

Fatalism, 37, 40, 44, 45, 46, 78

Fear, as motive for myth-making, 69–70

Ferré, Frederick, 210, 254

Festugière, 222

Fire, 266–267; in "Little Gidding" (Eliot), 256–257, 272–273

Fisher, R. A., 153, 159, 160

Four Quartets, "Little Gidding," (Eliot), 256–257, 272–273

France, scientific thought in, 14–15, 115, 121, 140–142, 147, 151–152, 156–159. *See also* Cartesianism; Jacob, François; Monod, Jacques, Teilhard de chardin, Pierre

Future of Man, The (Teilhard de Chardin), 114, 119

Future of the universe. *See* Apocalyse myths; Running-down universe

Galen, 227

Galileo, 3, 11, 95, 176, 219, 221; *Sidereus Nuncius* (*The Starry Messenger*), 219–220

Galton, Francis, 110; *Hereditary Genius*, 202

Garaudy, Roger, 141

Genesis, Book of, 2, 34

Ghost in the Machine, The (Koestler), 133–139, 177

Gibbons, Stella, *Christmas at Cold Comfort Farm*, 124

Gibbs, Josiah Willard, 160

God That Failed, The (Koestler), 177

Goethe, Johann von, 109, 130, 131, 137, 187

Goldiamond, Israel, 182

Gravitational force: Kant on, 4–5;

Newton's theory of, 28–32

Greene, Graham, 114

Grene, Marjorie, 109

Gustafson, James, 268, 269–270, 271

Gutenberg, Johannes, 94, 98

Haeckel, Ernst, 12; *Riddle of the Universe*, 10

Haldane, J. B. S., 159, 160

Hales, Stephen, 238

Halley, Edmund, 259

Hardy, Alister, 184

Harvey, William, 95, 227

Hasard et la Nécessite, Le (Monod), 141–155, 156, 157, 195

Haydn, Joseph, *The Creation*, 217–218

Hegel, Georg W., 144, 167

Hegelian theorists, and evolution, 63, 64

Heidegger, Martin, 238, 256–257

Heilbroner, Robert, 172

Heisenberg, Werner K., 42, 249

Heraclitus, 2, 144, 167

Herder, Johann G., 140

Hereditary Genius (Galton), 202

Hierarchical schemes, 54–55

Himmelfarb, Gertrude, 109

Holton, Gerald, 15

Hoyle, Fred, 35–36, 82, 83, 172

Hume, David, 5, 13, 123

Humor, Koestler on, 92–93

Hutchinson, Evelyn, 265, 266

Huxley, Aldous, 68

Huxley, Julian, 14, 114, 116, 128; on evolution and ethics, 13, 56–68, 129

Huxley, Thomas Henry, 60, 64–65, 67, 68, 129, 138, 171

Impossibilities, theoretical versus practical, 44–47

Inge, W. R., 35

In Memoriam (Tennyson), 146, 195n

Insight, Koestler on, 95, 96–97

Insight and Outlook (Koestler), 109

In the Days of the Comet (Wells), 38

Institut Pasteur, 156, 157

Isaiah, 222

Jacob, Francois, 14, 15; *La Logique du*

Vivant, 156–164

Janus: A Summing Up (Koestler), 178, 179–180, 208

Jeans, James H., 73

Jonas, Hans, 251

Jung, Carl, 99

Kammerer, Paul, *Das Gesertz der Serie*, 178

Kant, Immanuel, 8, 9, 13, 151, 189, 234; *Allegemaine Naturgeschichte und Theorie des Hummels*, 4; cosmology of, 3, 4–8; *Critique of Pure Reason*, 4, 6, 9, 150; *Die Ende Aller Dinger*, 4; ethics of, 6, 250; and Newton's theories, 4–6, transcendental philosophy of, 3, 4, 6–8

Kekulé von Stradnitz, Friedrich, 97

Kendrew, John, 157

Kepler, Johann, 82, 219; Koestler on, 90, 94–95, 109, 132, 176, 178, 179

Kidinnu, 222, 223, 225

Kierkegaard, Søren, 152

Kinesics, 204

Koestler, Arthur, 14, 89–113, 127–139, 146, 149, 176–200; *The Act of Creation*, 90–113, 177; arguments of, evaluated, 91–111, 132–138, 180–196, 198–199; *The Case of the Midwife Toad*, 178; *Darkness at Noon*, 90, 139, 177, 197; *The Ghost in the Machine*, 132, 133–139, 177, 184; *The God That Failed*, 177; *Insight and Outlook*, 90, 109; *Janus: A Summing Up*, 178, 179–180, 208; on neo-Darwinist evolution theory, 109–110, 181, 183–185, 190–191, 195–196; on psychology, 133–134; as scientist, 91; *Scum of the Earth*, 177; *The Sleepwalkers*, 90, 109, 176–177, 178; *Thieves in the Night*, 177

Köhler, Wolfgang, 95

Koyre, Alexandre, *Newtonian Studies*, 240

Kubla Khan (Coleridge), 97

Kuhn, Thomas, *The Structure of Scientific Revolutions*, 211

Küng, Hans, 268, 269

Lamarck, Jean B., 14, 120, 137, 140

Lamarckian evolution theory, 14, 120, 121—122, 137, 140, 177, 184, 190
Laplace, Pierre Simon, 5, 13, 209, 243, 248, 249
Laurence, William, *Lectures on Physiology, Zoology and the Natural History of Man*, 166
Lavoisier, Antoine, 100—101
Leavis, F. R., 172
Lectures on Physiology, Zoology and the Natural History of Man (Laurence), 166
Leibniz, Gottfried von, 16, 30, 130, 193—194, 274; compared with Koestler, 196—197, 198; *Monadology and Theodicy*, 197; organicism of, 131—132, 138
Leopardi, Giacomo, 77, 80
Levins, Richard, 159
Levi-Strauss, Claude, 172
Lewontin, Richard, 159, 191
Limbic system, 135, 169, 199
Linguistics, 148, 162, 163, 175, 204
"Little Gidding" (Eliot), 256—257, 272—273
Locke, John, 141
Loeb, Jacques, 149
Logique du Vivent, La (Jacob), 156—163
Luckman, Thomas, *The Structure of Scientific Revolutions*, 211
Luria, A. R., 133, 141
Lysenko, Trofim D., 122

McCarthy, Mary, 140
Mach, Ernst, 154
Mahabharata, 2
Malthus, Thomas R., 160
Maritain, Jacques, 153
Marx, Karl, 144, 167
Marxist theorists, and evolution, 63
Massachusetts Institute of Technology, 111
Mathematical Principles of Natural Philosophy (Newton), 6
Maupertius, Pierre Louis Moreau de, 4
Maxwell, James C., 160
Mayr, Ernst, 159, 191
Mead, Margaret, 202
Mechanism, Cartesian, 14—15, 111, 130—131, 147, 160, 161

Medawar, Peter, 150n
Megiste Syntaxis (Ptolemy), 10, 223, 224
Mendel, Gregor J., 158
Mercator's projection, 46, 47
Merleau-Ponty, Maurice, 153
Metaphysics (Aristotle), 150
Methodology, Cartesian, 130—137, 140, 209, 210
Meyerson, Émile, 151
Milne, John, 51
Mind and Nature: A Necessary Unity (Bateson), 205—209, 210—213
Molecular biology, 157—158, 160—161
Monad, Jacques, 14, 15; on evolution theory, 14, 141, 143, 144, 146, 147, 152; *Le Hasard et la Nécessité* (Chance and Necessity), 141—155, 156, 157, 195; natural philosophy of, 141—142
Monadology and Theodicy (Leibniz), 197
Montaigne, Michel de, 3, 126, 138, 172
Moore, G. E., 73
Morality. *See* Ethics
Morgan, T. H., 157
Morris, Desmond, *The Naked Ape*, 182
Moses, as historian, 218
Mountolive (Durrell), 222—223
Muir, John, 265—266
Mutations, 143, 145; random, 184—185
Myths: Greek, 23, 24; justificatory purposes of, 53; mechanomorphic, 24; motives for creating, 23, 53, 69—70; Near Eastern, 25; and philosophy, 9; twentieth century, 23—24. *See also* Scientific myths

Naked Ape, The (Morris), 182
National Science Foundation, 127
Natural religion, 218, 221; reunion with, 217, 218—219, 221, 255—256, 261
Natural selection, 94, 121, 122, 123, 145, 147, 184. *See also* Evolution; Evolution theory, Darwinian
Natural Theology (Paley), 66, 67

Nature (weekly), 13
Needham, J. S., 79
Neocortex, 135, 169, 199
Neo-Darwinism, 145, 146, 153, 157, 190−191; Koestler on, 109−110, 181, 183−185, 195−196
Neugebauer, Otto, 223
Newton, Isaac, 5, 6, 9, 16, 84, 95, 130, 218; Halley on, 258−259; introduction of theory of gravity, 28−31; and theology, 193, 232; *Principia Mathematica Philosophiae Naturalis*, 6, 217, 219
Newtonian Studies (Koyre), 240
Nietzsche, Friedrich, 152

Objectivity, scientific, 211−213, 237−253, 256
Omniscient Calculator, Laplace's image of, 243, 248, 249
Optimism, and world view of biologists, 78
Origin of Species, The (Darwin), 146, 165, 218
Origins of Modern Science, The (Butterfield), 254
Orthogenesis, 145−146
Osiander, Andreas, 10−11
Ostwald, F. W., 35, 36, 79, 80
Our Knowledge of the External World (Russell), 211

Paley, William, 66, 67
Paralogisms, 7
Parsons, Talcott, 207
Pascal, Blaise, 89, 224
Pauling, Linus, 230
Pavlov, Ivan Petrovich, 108, 133
Peguy, Charles, 153
Peirce, Charles Sanders, 154, 212
Pepys, Samuel, 127
Pessimism, and world view of physicists, 77−78
Phénomene Humaine, Le (Teilhard de Chardin), 114, 116−117, 123, 125
Philosophers: Aristotole on, 238; on concept of universe as a whole, 2, 3; Greek, 2−3, 9; mathematical, 3; synthetic, aims of, 73−74
Philosophical Investigations (Wittgenstein), 150

Philosophy: and mythology, 9; positivist, 17, 83, 84; as separated from science, 72; synthesis with science, 72−74, 80
Physical cosmology of Kant, 4−7
Physical theories, in solution of ethical or political problems, 32
Physicists: atomistic concepts of, 9; and scope or physical science, 2; world views of, 75−78
Physics: Kant's writings on, 4; misapplication of theories of, 38−42
Physique d'un Croyant (Duhem), 11, 115
Plato, 2−3, 6, 143; *Republic*, 224; *Timaeus*, 25, 222
Platonists, 167
Poincaré, Jules Henri, 96, 151, 226
Polanyi, Michael, 109, 128, 149
Popper, Karl, 80; *Conjectures and Refutations*, 127
Poseidon, 23, 24, 70
Positivists, 17, 83, 84
Postmodern science, 210−211, 213; Bateson and, 210−211, 213; defined, 210, 254; world view of, 255−257, 262
Predictions: of fate of universe, 34−36 (*see also* Apocalypse myths; Running-down universe); of scientists, 34−35
Pribram, Karl, 141
Priestly, Joseph, 16, 233
Principia Ethica (Spencer), 73
Principia Mathematica Philosophiae Naturalis (Newton), 217, 219
Professional Foul (Stoppard), 238
Psychological evolution, 203−205, 208−209
Psychological theory, Koestler on, 103, 104−108, 133−134
Psychology: behaviorist, 181−183; Darwin and, 166
Psychopharmacology, 135, 139, 199−200
Psychotherapy, and philosophy, 262
Ptolemaic cosmology, versus Copernican theory, 81−82, 227
Ptolemy, Claudius, 176; *Almagest*, 10; *Megiste Syntaxis*, 10, 223, 224; *Tetrabiblos*, 225

Pythagoras, 9

Rabut, Olivier, 122, 124
Ramsey, Frank, 36
Randomness, 145, 184–185
Ray, John, 232; *The Wisdom of God, as Manifested in the Works of His Creation*, 218
Renaissance, scientific discovery during, 176, 178
Renan, Ernest, 153
Republic (Plato), 224
Revel, Jean-François, 140
Riddle of the Universe (Haeckel), 10
Rosencrantz and Guildenstern Are Dead (Stoppard), 237, 242
Royal Society of London, 241
Running-down universe, 35–45, 47, 48–49, 78; as scientific myth, 22, 26–27, 49
Russell, Bertrand, 207, 211–212; *The Analysis of Matter*, 212; *Our Knowledge of the External World*, 211

Sacred History of the Earth (Burnet), 9
Sagan, Carl, 2, 14, 165–175; *The Dragons of Eden: Speculations on the Origins of Human Intelligence*, 168–175
Sahlin, Marshall, 168, 172, 174
Sanctions, and cosmology, 53. *See also* Ethics
Scale of Nature. *See* Sovereign Order of Nature
Schiller, Friedrich von, 131
Schiller, Johann, 187
Schopenhauer, Arthur, 152
Schrödinger, Erwin, 230
Science: application of concepts of, 8–9, 10, 251–252; atomistic concepts of, 7, 9; increasing specialization of, 230, 233; and natural religion, 12–13, 217–219, 221, 227, 231, 255–256, 261; popular, 21–22, 85; positivist philosophy of, 15; as separated from philosophy, 72–74, 80; subdivision of natural world by, 7–8; synthesis with philosophy, 72–74, 80
Scientific concepts: Koestler on evolu-

tion of, 94–97; risk of overinterpretation of, 8–9
Scientific mythology, 14, 19–85; Evolutionary Ethics as, 58–70; as justification for ethical and political views, 53–70; mechanomorphic nature, 24; recognition of, 24–25; versus scientific theories, 21–32, 38–52; world views of scientists, 72–80. *See also* Running-down universe
Scientific terminology, 25, 27–31; ambiguity in use of, 26; abstract versus concrete, 44; classification of chemical substances, 45–46; nonscientific uses of, 27–32
Scientists: Aristotle on, 238–239; constitutional rights of, 251; as differentiated from artists, 110–111; involvement of, 251–253, 255, 256, 267; looked to for guidance, 81–83; objectivity of 211–213, 237–253, 256; as ordinary men, 81, 83; Protestant, 218, 232; spectator role of, 237–253, 256. *See also* Biologists; Physicists
Scum of the Earth (Koestler), 177
Second Law of Thermodynamics. *See* Thermodynamics, Second Law of
Sedgwick, Adam, 122, 146n–147n, 208
Shakespeare, William, *Hamlet*, 237
Shaw, George Bernard, 121, 194
Sidereus Nuncius (Galileo), 219–220
Sierra Club, 266
Simmel, Georg, 154
Simpson, George Gaylord, 116, 159
Skinner, B. F., 90, 133, 181, 183; *Beyond Freedom and Dignity*, 182
Sleepwalkers, The (Koestler), 109, 176–177, 178
Smith, Adam, 230
Snow, C. P., 173
Social Construction of Reality, The (Berger and Luckman), 211
Social structure, mythical foundations for, 53
Sociobiology (Wilson), 165, 168
Sociocultural change, and evolution, 163
Socrates, 2, 10, 130, 172

Solar system, 29, 31, 82. *See also* Astronomy

Sovereign Order of Nature, as scientific mythology, 54—56

Space, 1; notions of, 6—7

Space travel, 168

Spectators, scientists, as, 237—253, 256

Spencer, Herbert, 121, 141, 154; *Principia Ethica*, 73

Spencer-Jones, Harold, 40, 41, 49—50, 51

Spinoza, Baruch, 124

Steps to an Ecology of Mind (Bateson), 205

Stoic position, 222, 263, 264

Stoppard, 253; *Professional Foul*, 238; *The Real Inspector Hound*, 237—238, 243; *Rosencrantz and Guildenstern Are Dead*, 237, 243

Structuralism, in biology, 161—162, 164

Structure of Scientific Revolutions, The, 211

Synthesis, of science with philosophy, 72—74, 80

Taine, Hippolyte Adolphe, 153

Teilhard de Chardin, Pierre, 2, 14, 15, 128, 137, 149; *The Future of Man*, 114, 119; *Le Phénomene Humaine* (The Phenomenon of Man), 2, 114, 116—117, 123, 125; training of, 115—116; view of evolution, 119—120; world view of, 117—119, 124, 125—126

Tennyson, Alfred, *In Memoriam*, 146, 195n

Terms, scientific. *See* Scientific terminology

Tetrabiblos (Ptolemy), 225

Thales, 2, 9, 172

Theology: Catholic, 116; in eighteenth century, 21; in relation to science, 8, 12—13, 221, 227—231, 255—256, 261; in twentieth century, 21, 254, 255—261, 268—274

Theoretical attitude, 238—239

Theoretical impossibilities, 44—47

Theory, 238—240; etymology of term, 239; philosophy as, 238—239, 240

Thermally-isolated system, universe as, 37—38, 40, 42—43

Thermodynamics, Second Law of, 26, 27, 47, 48—49; apocalyptic implications of, 35—41, 47—49; as a law about the universe, 47, 48—49; and running-down universe, 35, 36—42, 47—49; as universal law, 40—42

Thieves in the Night (Koestler), 177

Thucydides, 167

Timaeus (Plato), 25, 222

Time: beginning of, 49—50, 51; concepts of, 1, 6—7, 49—50, 51

Time-scales, 51, 52

Titchener, E. B., 183

To the Lighthouse (Woolf), 70—71

Toynbee, Arnold, 114

"Transcendental" method of Kant, 3, 4, 6—8

Ullstein press, 177

Universal History of Nature (Kant), 4

Universal laws, 40—41, 84

Universe: arbitrary subdivisions of, 7—8; Kant on, 3—7; knowledge of, 3—4; laws about, 48; physical versus biological views of, 72—78; as a whole, 1—2, 6, 7, 8

Velikovsky, Immanuel, 9

Vico, Giambattista, 161, 167, 263

Vygotsky, L. S., 133, 141

Waddington, C. H., 73

Washburn, Sherwood, 173

Watson, James D., 157, 191

Watson, John B., 133, 181, 183

Weber, Max, 230, 235

Weismann, Auguste, 147

Wells, H. G., 80; *In the Days of the Comet*, 38

Wenner-Gren Foundation, 116

Wheeler, John, 2, 13

Whiston, William, 9

Whitehead, Alfred North, 144

Wilson, Edward O., 173—174, 191; *Sociobiology*, 165, 168

Wisdom of God, as Manifested in the Works of His Creation, The (Ray), 218

Wittgenstein, Ludwig, 13, 14, 16; *Philosophical Investigations*, 150
Woolf, Virginia, *To the Lighthouse*, 70—71
World views: mechanicist, 130—132; of scientific disciplines, 72—78
World wars, 197
Wotan, 23, 24, 76

Wright, Sewall, 153, 159, 160
Wundt, Wilhelm, 183

Young, J. Z., 82, 83

Zeus, 23—24
Zoroaster, 222